# Basic Principles of Plate Theory

Published by Surrey University Press
A member of the Blackie Group
Bishopbriggs, Glasgow and
Furnival House, 14–18 High Holborn, London


First published 1982



**British Library Cataloguing in Publication Data**

Lowe, P.G.
Basic principles of plate theory.
1. Structures, Theory of
I. Title
624.1′7 TA645

ISBN 0-903384-26-4
0-903384-25-6 Pbk



Printed in Great Britain by
Thomson Litho Ltd., East Kilbride, Scotland

# Basic Principles of Plate Theory

**P. G. Lowe**
Ph.D., M.I.C.E.

Professor of Engineering
University of Auckland
New Zealand

formerly

Professor of Structural Engineering
University of Strathclyde
Glasgow, U.K.

**Surrey University Press**

*To Alexander Robert Dykes*
*1927–1980*
*Friend and Colleague*

# Preface

Adding another volume, even if only a slim one, to the technical books already published requires some justification. Mine is, firstly, that plate theory is not well represented in the available elementary texts, and secondly that no existing text adequately covers modern applications. The present account is intended to be elementary (though this is a relative term) while still providing stimulation and worthwhile experience for the reader. Special features of interest will I hope be the treatment of geometry of surfaces and the attempts around the end of the work to speculate a little. The detailed treatment of geometry of surfaces has been placed in an appendix where it can readily be referred to by the reader.

My interest in plate theory extends back many years to the energetic and stimulating discussions with my supervisor, Professor R. W. Tiffen, at Birkbeck College, London, and a debt to him remains. Interest was rekindled for me by Dr R. E. Melchers when I supervised him in Cambridge some ten years ago, and more recently my stay at Strathclyde University and encouragement and stimulation in the Civil Engineering Department led me to undertake the present work.

The typescript was prepared by Ms Catherine Drummond and I thank her warmly for this and other assistance, always cheerfully offered. My thanks also to the publishers and the referees for useful comments and advice.

P.G.L.

# Notation

| | |
|---|---|
| ═══════ | Simply supported edge |
| × × × × × × × | Fixed (clamped) edge |
| ——————— | Free edge |
| $(\ ).(\ )$ | Scalar product |
| $(\ )\times(\ )$ | Vector product |
| A | Slab area |
| B | Boundary length, perimeter |
| D | Plate stiffness |
| E | Young's Modulus (the context will prevent confusion with next symbol) |
| E, F, G | Coefficients of first fundamental form |
| L, M, N | Coefficients of second fundamental form |
| $\mathbf{i}, \mathbf{j}, \mathbf{k}$ | Cartesian unit vectors |
| $\mathbf{I}, \mathbf{J}$ | Polar unit vectors |
| k | Constant curvature |
| L | Span, beam or plate |
| $M_p$ | Full plastic moment |
| $M_{\alpha\beta}$ | Moment stress resultants |
| $Q_\alpha$ | Transverse-force stress resultants |
| $\mathbf{n}$ | Unit normal vector |
| p | pressure loading |
| $p_L, p_c, p_u$ | Collapse pressures |
| $r, \theta$ | Polar coordinates |
| $\mathbf{r}$ | Radius (or position) vector |
| s | Arc length |
| $(\ )_s$ | Derivative with respect to s |
| t | Plate thickness |
| u, v | Surface coordinates |
| $(\ )_u, (\ )_v$ | Derivatives with respect to u, v |
| w | Transverse displacement |
| x, y, z | Cartesian coordinates |

| | |
|---|---|
| $(\,.\,.\,)_{,x}$ | Derivative with respect to x |
| $\alpha$ | Parameter = $B^2/A$ |
| $\beta$ | Parameter = $\gamma/\alpha$ |
| $\gamma$ | Parameter = $p_c A/M_p$ |
| $\kappa$ | Curvature |
| $\boldsymbol{\kappa}$ | Curvature matrix |
| $\nu$ | Poisson's ratio |
| $(\ )_L$ | Lower bound collapse value |
| $(\ )_c$ | Exact collapse value |
| $(\ )_u$ | Upper bound collapse value |
| $(\ )^{IV}$ | Fourth derivative, with respect to x, or according to context. Also $(\ )^{I}$, $(\ )^{II}$, $(\ )^{III}$ used |

# Contents

## Chapter 4. PLASTIC PLATES



## Chapter 5. OPTIMAL PLATES



## Chapter 6. BIBLIOGRAPHY AND EXERCISES



## Appendix GEOMETRY OF SURFACES

# 1 Preliminaries

## 1.0 Motivation

The topic to be studied is the mechanics of plates. First, what is a plate? Here a plate will be defined to be a plane, load-carrying element in a structure which is comparatively thin in comparison with the plate surface dimensions. Typical examples of plates in practical applications are floor slabs in buildings, decking on bridges, load-bearing wall structures, ship hull plating and many other constructions. Plate elements may be large, spanning tens of metres in buildings or bridges, or very small when incorporated as a diaphragm in a pressure reduction valve. In all cases however the elements have thicknesses which are a tenth to a twentieth or less of the radius of the valve diaphragm, or span of the slab.

Note that the emphasis will be placed on the *mechanics* of plates. Mechanics is an old and honourable branch of quantitative scientific and technical study. There are various aspects to the study.

First, there is the need for some technical apparatus to undertake the study. Most of this apparatus will be reviewed or set up in this present chapter. Secondly, there is the strict mechanics content of the subject, the equations of statics (or dynamics) suitably written to describe the essentially two-dimensional plan form of the plate and the actions induced in it. Along with the equations of statics there is a central need to study the *geometry of deformation* of the plate. This study is probably best thought of as a third study, related to though separate from the statical content.

Fourthly, there will arise the question of modelling the physical materials from which the plates are made. Usually metal plates respond in an elastic manner for the lower load requirements but for higher load regimes they respond plastically. The elastic response is a proportional, linear response; the plastic is a limiting behaviour displayed by ductile materials such as steel. Many plates in practical cases are constructed from reinforced (or perhaps prestressed) concrete. Such a material is a composite of concrete matrix with embedded steel reinforcement.

The technical requirement is to produce a composite with ductile metal-like qualities while using one component (concrete) which is far from ductile. The development of elastic and plastic themes will occupy the major part of the whole work, since under these heads will be discussed the most important technical problems of plates.

In addition to the conventional theory of plates, as covered by the topics discussed above, a different and quite new type of plate theory will be outlined in the later part of the work. The technical importance of this new and in some ways unusual theory is not yet established. However, it is felt desirable to introduce such a discussion at a relatively early stage in order to demonstrate the power and scope of the basic ideas.

Our subject has a relatively short history if a strict, matter-of-fact attitude is adopted. Only in the last hundred, or possibly the last 50 years have certain of the technical problems in plates become soluble in modern terms. However, in important respects the subject has roots which extend back at least three hundred years and some appreciation of this heritage together with the more recent and accessible literature on the subject are dealt with in the closing chapter.

The purpose of this work is to encourage the reader to acquire some knowledge of, and gain some expertise to tackle and solve, a range of problems in the field of technical plate theory. Exercises are provided at various points, and it is strongly recommended to the reader that these exercises be attempted as the various sections and chapters are studied and mastered. Only by regular problem solving can many of the important points of principle and detail be appreciated.

## 1.1 Vectors—algebra

At various points in the later chapters, but especially in connection with the study of the relevant geometrical content of the subject, it will be convenient to offer a vector formulation. It is proposed therefore to review those aspects of vectors which will be drawn upon later.

A vector is a quantity which possesses both *magnitude* and *direction*. Examples of vectors are forces and surface slopes. There is an important additional class of vectors with which the geometrical part of the subject will deal—the class of *position vectors*.

Visualize a coordinate system, say a rectangular cartesian system of straight, orthogonal axes (Fig. 1.0). Any point such as P(x, y, z) when referred to the origin O, for example, will be spoken of as possessing a *position*, or *radius*, vector $\mathbf{r}$ as shown in the figure.

Once the origin has been chosen and the point P selected then the position

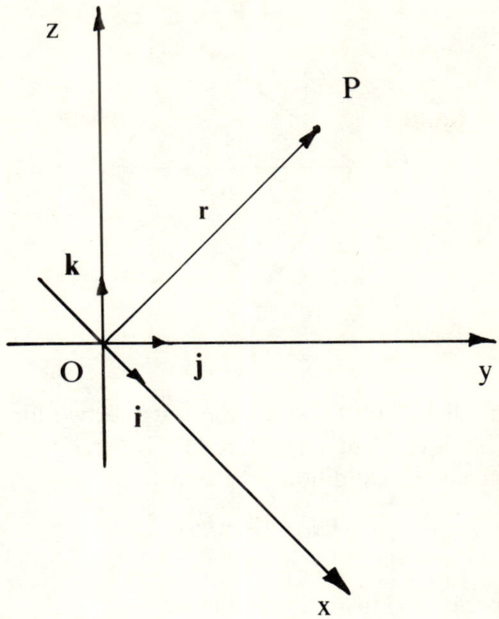


**Figure 1.0** The position vector.

vector **OP** will be fixed, and will have the attributes of a vector, namely of magnitude and direction.

When a vector has the special quality that its magnitude is fixed and equal to unity, then that vector will be called a *unit vector*. This is the defining quality; a unit vector has unit magnitude. There is no restriction placed on the direction of the unit vector.

Associated with the rectangular cartesian axes x, y, z (Fig. 1.0), it is conventional to define three unit vectors **i**, **j**, **k**, directed in turn along the axes x, y and z. Thus **i** is by definition the unit vector directed along the x axis, **j** along the y and **k** along the z axis. In this case, these three special unit vectors *do* have restricted directions—those associated with x, y and z respectively.

The position vector **r** of point P, with respect to O, can be written as

$$\mathbf{r} = x\mathbf{i} + y\mathbf{j} + z\mathbf{k} \tag{1.1.0}$$

where now x, y, z are the *coordinates* of P. Alternatively, x, y and z can be called the *components* of **r** with respect to the axes x, y, z. They are the lengths of the sides of the rectangular parallelapiped with OP as the diagonal.

Vectors may be added and subtracted, according to the parallelogram rule. If two forces acting at a common point are added together, a *single* equivalent force, often known as the *resultant-force*, is produced. The resultant is the

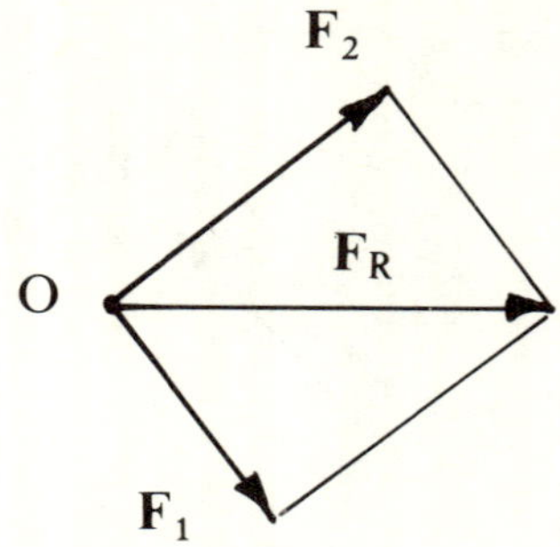


**Figure 1.1** Vector addition.

*diagonal* of the parallelogram of which the original two forces are adjacent sides (Fig. 1.1). This equivalent single force $\mathbf{F}_R$ is correct for magnitude and direction. Algebraically this addition is written

$$\begin{aligned}\mathbf{F}_R &= \mathbf{F}_1+\mathbf{F}_2\\ &= \mathbf{F}_2+\mathbf{F}_1.\end{aligned} \tag{1.1.1}$$

Subtraction is performed in a similar manner:

$$\mathbf{F}_2 = \mathbf{F}_R-\mathbf{F}_1$$

or

$$\mathbf{F}_1 = \mathbf{F}_R-\mathbf{F}_2 \tag{1.1.2}$$

and the subtraction can be illustrated on a diagram where it is seen that $\mathbf{F}_2$ results from the addition of $\mathbf{F}_R$ and $(-\mathbf{F}_1)$ (Fig. 1.2).

This discussion of addition and subtraction of vectors has been couched in terms of force vectors, but the same rules and physical meanings can be attached to these operations applied to other sorts of vectors. The only requirement is that the vectors be of the *same type* for them to be combined, namely they must both be forces or radius (position) vectors, for example.

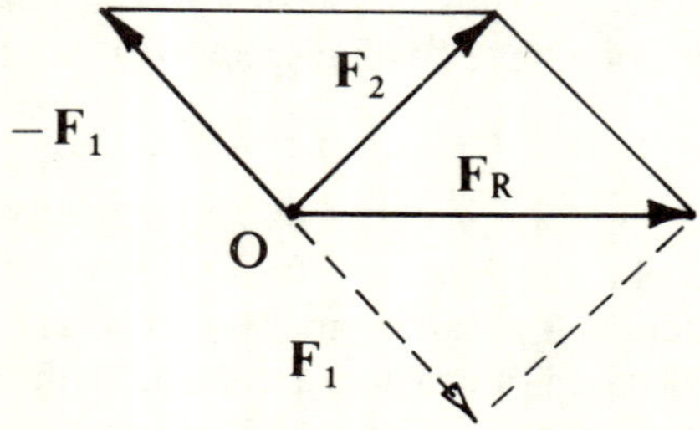


**Figure 1.2** Vector subtraction.

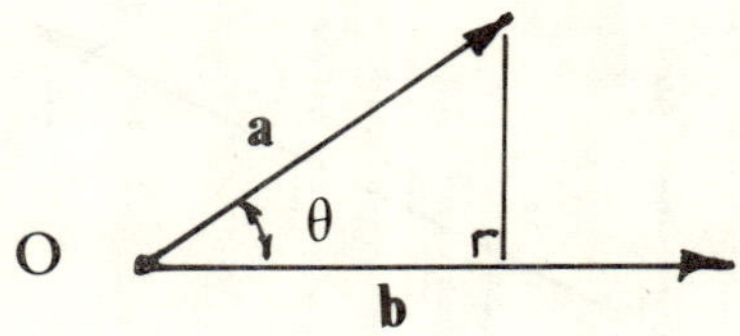


**Figure 1.3** Scalar product.

The algebra of vectors can be extended beyond addition and subtraction to include multiplication, but now there are *two* types of multiplication. First there is the operation of multiplying two vectors together so as to form a scalar—this is called *scalar multiplication*. Recall, too, that a scalar is an ordinary number, which has magnitude but *no* directional quality associated.

This operation of *scalar multiplication* of two vectors **a** and **b** is defined to be

$$\mathbf{a} \cdot \mathbf{b} = |\mathbf{a}|\,|\mathbf{b}| \cos \theta \qquad (1.1.3)$$

where $|\mathbf{a}|$ is the symbol used to indicate *magnitude* of **a** and $\theta$ is the angle between **a** and **b** (see Fig. 1.3). From the nature of the right-hand side of (1.1.3) it can be confirmed that **a** . **b**, the scalar product of **a** and **b**, is indeed a scalar. The reason for introducing this product is one of convenience. An interpretation is that **a** . **b** is the scalar value of the magnitude of **b** multiplied by the component of **a** in the direction of **b**. This latter quantity is the magnitude of the vector **a** thought of as composed of two components, the one *along* **b** and the other normal to **b** (see Fig. 1.3).

Thus

$$\mathbf{a} \cdot \mathbf{b} = \mathrm{b}(\mathrm{a} \cos \theta).$$

Alternatively, the roles of **a** and **b** can be exchanged and the result is unaffected. Hence **a** . **b** = **b** . **a**, where the scalar product is *associative*.

Return to consider the position vector **r** of (1.1.0). The most convenient manner for finding the components of **r**, given (1.1.0), is to form the scalar products **r** . **i**, **r** . **j**, **r** . **k** in turn. Then, because **j** and **k** are mutually perpendicular to **i**, their respective $\theta$'s are $\pi/2$ and hence

$$\mathbf{i} \cdot \mathbf{j} = 0 = \mathbf{i} \cdot \mathbf{k}$$

Hence

$$\mathbf{r} \cdot \mathbf{i} = \mathrm{x}\mathbf{i} \cdot \mathbf{i} + \mathrm{y}\mathbf{j} \cdot \mathbf{i} + \mathrm{z}\mathbf{k} \cdot \mathbf{i} \qquad (1.1.4)$$
$$= \mathrm{x}.$$

This result follows since **i** . **i** = 1. **i** being a unit vector and $\theta$ here is zero.

Relation (1.1.4) serves to illustrate a useful role of orthogonal unit vectors and the scalar product.

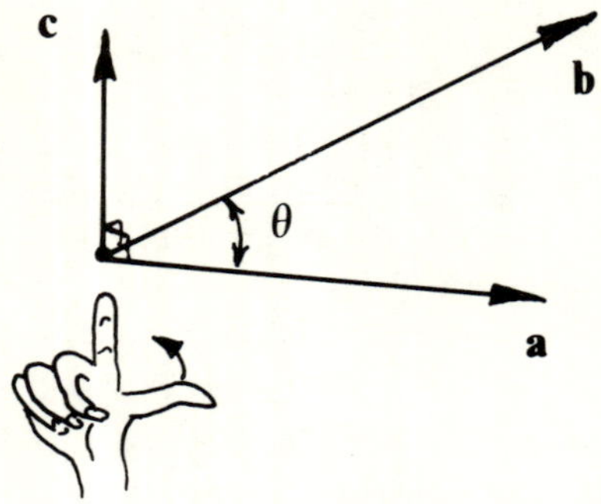


**Figure 1.4** Vector product.

*Exercise:* The magnitude of a vector, **r**, can conveniently be found by noting that **r**.**r** is a scalar and has magnitude $|\mathbf{r}|.|\mathbf{r}|\cos\theta = r^2$, where $|\mathbf{r}|$ is the magnitude of **r**. Hence show that the magnitude of **r** with x, y, z components (3, 4, 5) is $5\sqrt{2}$.

The second type of multiplication of vectors is the *vector product*, so called because the result of multiplying two vectors in this case is to produce a further vector. The operation is denoted by ( ) × ( ). Hence

$$\mathbf{a} \times \mathbf{b} = \mathbf{c}$$

where $|\mathbf{c}| = ab\sin\theta$ and has a direction given by a right-hand rule with respect to **a** and **b** (see Fig. 1.4).

Again, the motivation for introducing the vector product is one of utility.

Note now that if the vectors multiplied in a vector product are reversed, then the *direction* of the resulting vector is *reversed.* Hence the operation ( ) × ( ) is *not* commutative, that is the order of multiplication of terms cannot be reversed without changing the result.

*Exercise:* Consider the trio of fixed unit vectors **i**, **j**, **k**. Suppose x, y, z form a right-handed triad of axes. This means that applying the right-hand rule to carry x into y (by motion of the thumb) ensures that the forefinger is pointing along the positive z axis during the process.

Hence show that $\mathbf{i} \times \mathbf{j} = \mathbf{k}$, $\mathbf{k} \times \mathbf{j} = -\mathbf{i}$, $\mathbf{i} \times \mathbf{i} = 0$ and similar results.

## 1.2 Vectors—calculus

The discussion of vectors thus far has been concerned with *algebraic* operations on vectors. Note, too, that no operation of *division* has been defined.

There is use for a little of the calculus of vectors, particularly differentiation of vectors, in the following chapters.

A vector, such as a position vector (**r**) changes with choice of the point P. Hence **r** can be thought of as a function of the coordinates of P. As different points P are chosen, so **r** changes and hence rates of change and differentiation become natural developments.

Consider a particular circumstance of change. The polar coordinate system (Fig. 1.5) is specified by r, $\theta$. Associated, and defined are the unit vectors **I**, **J**. As r and $\theta$ change then **I** and **J** are likely to change, but by how much?

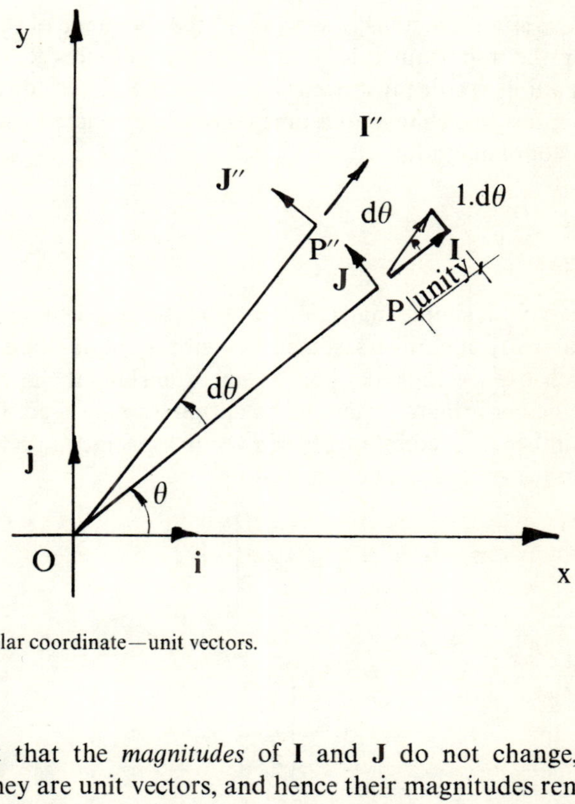


**Figure 1.5** Polar coordinate—unit vectors.

Note first that the *magnitudes* of **I** and **J** do not change, because by definition, they are unit vectors, and hence their magnitudes remain at unity. However the *directions* of **I**, **J** may change, and hence, so far as unit vectors are concerned, changes are limited to changes of direction.

In the present case, if **r** changes to **r**+d**r** but $\theta$ is held constant then P moves to P′ where PP′ = d**r**. However, the directions of **I** and **J**, the new unit vectors at P′ are unchanged, and hence **I**′ = **I**, **J**′ = **J**.

But should $\theta$ change to $\theta+\mathrm{d}\theta$ and P move to P″ (Fig. 1.5) then the directions of the associated **I**, **J** do change. From the geometry of the figure it is seen that

$$\mathbf{I}'' = \mathbf{I} + \mathrm{d}\theta \,.\, \mathbf{J}$$

and (1.2.0)

$$\mathbf{J}'' = \mathbf{J} - \mathrm{d}\theta \,.\, \mathbf{I}.$$

Adopting the notation of scalar differential calculus, then for $\mathrm{d}\theta$ an infinitesimal, $\mathbf{I}'' = \mathbf{I} + \mathrm{d}\mathbf{I}$ and

$$\frac{\mathrm{d}\mathbf{I}}{\mathrm{d}\theta} = \mathbf{J}, \quad \frac{\mathrm{d}\mathbf{J}}{\mathrm{d}\theta} = -\mathbf{I}. \tag{1.2.1}$$

Hence there arise differentials of vectors. In the use made of the concept of differentiation, the important role of unit vectors will emerge. In summary, any *change* in a unit vector must occur *normal* to the current direction of the unit vector, because the change to a unit vector is associated with change of direction, but not of magnitude.

## 1.3 Matrices

Some familiarity with simple matrix operations will be useful later. A matrix is a rectangular array of numbers, sometimes with equal numbers of rows and columns, which can be looked upon as an extension of the concept of a vector. Once the coordinate system to be adopted is specified, then a single list of three numbers (the components of a vector) specifies a vector. Thus the position vector **r** of earlier can be specified by

$$\mathbf{r} = (\mathrm{x}, \mathrm{y}, \mathrm{z}) = \begin{bmatrix} \mathrm{x} \\ \mathrm{y} \\ \mathrm{z} \end{bmatrix}.$$

Another notation is

$$\mathrm{r_i} = \begin{bmatrix} \mathrm{x} \\ \mathrm{y} \\ \mathrm{z} \end{bmatrix} \begin{matrix} 1 \\ 2 \\ 3 \end{matrix}$$

where the "free" subscript i can take on the value 1, 2 or 3 to identify the x, y or z component. Such a notation emphasises the *single subscript* aspect of a vector. Vectors, too, can have more than three components, though interest here is limited to three component vectors.

In contrast, matrices are *double subscript* objects, since two subscripts are needed to specify first the row and second the column which a particular element (or component) occupies. Of special interest are the simplest matrices which consist of just two rows and two columns and hence four elements or components at most.

Consider such an array

$$\mathbf{A} \equiv \begin{bmatrix} a_{11} & a_{12} \\ a_{21} & a_{22} \end{bmatrix} = a_{ij}. \tag{1.3.0}$$

Notation can be confusing. As far as possible, **bold** face upper case letters will be used to represent *matrices*, the size of the matrix (the number of rows/columns) being indicated either explicitly or from the context. In contrast, lower case bold face letters will be reserved for *vectors*.

Should it happen that $a_{12}$ equals $a_{21}$ in (1.2.2), then the matrix **A** is said to be symmetric. Most physical examples of matrices are symmetric matrices. Like vectors, matrices possess a non-commutative aspect in multiplication but they are associative in addition and subtraction. To be added or subtracted, matrices must be of similar shape—the same numbers of rows and columns—when the result of addition is that similarly positioned elements in the matrices are added to produce a new, similarly positioned element in the resulting matrix.

Matrix multiplication is achieved by taking a row in the first matrix and multiplying this, element by element, with a column of the second matrix, and summing the terms so found. The resulting new element is positioned in the result matrix in the position with the original row and column numbers of the row and column being multiplied .

It is necessary therefore that for two matrices to be multiplied they must be *conformable*, that is, have the same number of columns in the first as there are rows in the second. The resulting matrix will have a "size" given by the rows of the first and the columns of the second. Many important physical cases are concerned with square matrices and hence with equal numbers of rows and columns, when the above rules are much simplified.

*Example:* Given

$$\mathbf{A} = {}_{(2)}\overset{(3)}{\begin{bmatrix} 1 & 2 & 3 \\ 4 & 5 & 6 \end{bmatrix}},$$

$$\mathbf{B} = {}_{(3)}\overset{(2)}{\begin{bmatrix} 1 & 2 \\ 3 & 4 \\ 5 & 6 \end{bmatrix}},$$

Find $\mathbf{C} = \mathbf{A}.\mathbf{B} = c_{ij}.\ (i = 1, 2,\ j = 1, 2)$.

Consider

$$c_{11} = \underbrace{\overbrace{1 \times 1 + 2 \times 3 + 3 \times 5}^{\text{first row of A}}}_{\text{first column of B}} = 22,$$

$$c_{12} = 1 \times 2 + 2 \times 4 + 3 \times 6 = 28,$$

$$c_{21} = 4 \times 1 + 5 \times 3 + 6 \times 5 = 49,$$

$$c_{22} = 4 \times 2 + 5 \times 4 + 6 \times 6 = 64.$$

$$\therefore \mathbf{C} = \begin{bmatrix} 22 & 28 \\ 49 & 64 \end{bmatrix}$$

## 1.4 Statics—equilibrium

Newton's second law of motion for a particle of mass $m$ states that the rate of change of momentum is proportional to the force acting on the particle, and, as this is a vector equation, the direction of the velocity increase (the acceleration) is the same as that of the force acting. The main interest in the study of plates to follow is of equilibrium. Hence there are no accelerations, and as it will be assumed that the body was initially at rest, it therefore remains at rest, the resultant force acting on any particle must be zero, and the problem is one of statics.

Particles as such are seldom encountered in structural mechanics, and the concept of a body larger than a particle is therefore needed. Then the relevant law of statics is that the *resultant force* on this body must be zero, and the *resultant moment* of all the forces acting on the body about an arbitrary point must also be zero.

Generally, the moment of a given force about a chosen point changes with the choice of point. However, statics is concerned with a collection of forces acting on a body, and this system of forces has a zero resultant because the body is at rest. As the result of this, the resultant moment, made up of the vector addition of the moment of each individual force about the chosen point, if zero for *one* point will be zero for *all* points. Expressed in slightly different terms, if a system of forces has the property that the resultant moment of these forces is the *same* for all choices of point about which moments are taken, then the force system is equivalent to a *couple*—that is two equal and opposite forces not acting in the same line. If, in addition, the moment so calculated is zero, then the force system is equivalent to two equal and opposite forces but now acting collinearly. Such a force system is an *equilibrium* system. The words "statics" and "equilibrium" are used as

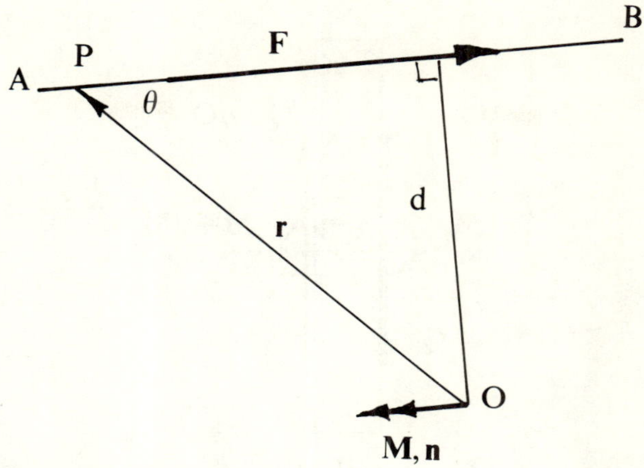


**Figure 1.6** Moment of a force.

alternatives but should really be used together to describe *static equilibrium*—the system being at rest in equilibrium.

The general definition of the moment (**M**) of a force (**F**) about a point O is that

$$\mathbf{M} = \mathbf{r} \times \mathbf{F} \tag{1.4.0}$$

where **r** is *any* radius vector joining O to a point on the line of action of **F**. From the definition of vector product, **M** is a vector, normal to the plane of **r** and **F**. The moment is the **turning effect** of **F** about O.

All the examples presently of interest are two-dimensional in the sense that all **F** and **r** lie in a common plane. Hence all the **M** vectors will be *parallel* and can be treated as scalars, to be added or subtracted. From Fig. 1.6 it is seen that

$$\begin{aligned} \mathbf{M} &= \mathbf{r} \times \mathbf{F} \\ &= (\mathrm{r}\,.\,\mathrm{F}\,.\,\sin\theta)\mathbf{n} \\ &= (\mathrm{F}\,.\,\mathrm{d})\mathbf{n} \end{aligned} \tag{1.4.1}$$

since $$\mathrm{d} = \mathrm{r}\sin\theta.$$

Here **n** is a unit vector normal to the plane of **r** and **F**. If all the **r** and **F** vectors lie in one plane, then all the **M** vectors will be parallel to **n**, and can be simply added or substracted, with **n** as a common factor.

The conditions for static equilibrium are then that

$$\mathbf{F}_{\mathrm{R}} = 0, \quad \mathbf{M}_{\mathrm{R}} = 0 \tag{1.4.2}$$

where $\mathbf{F}_{\mathrm{R}}$ and $\mathbf{M}_{\mathrm{R}}$ are, respectively, the *resultant force* and *resultant moment* of the individual forces about an arbitrary (i.e. any) point.

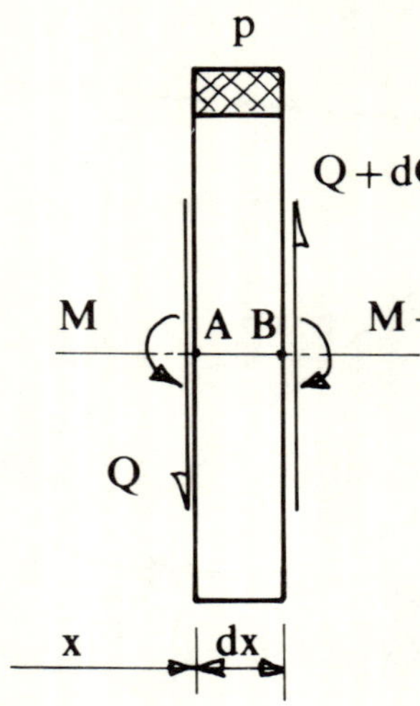


**Figure 1.7** Beam element.

*Exercise:* Assume a right-handed cartesian coordinate system and associated unit vectors **i**, **j**, **k**. A force **F** with components (1, 1, 1) acts through the point (0, 1, 0). Show that the moment (**M**) of **F** about point P with coordinates (1, 0, 0) is given by a vector with components (1, 1, −2). Show too that the axis of the moment **M** passes through R(2, 1, −2) and that PR is normal to the plane of **F** and P. Check that $M_p = \sqrt{6}$, PR $= \sqrt{6}$, QR $= \sqrt{8}$.

## 1.5 Summation convention and index notation

When multiplying vectors or matrices, many number pairs are generated and summed to find the new elements. If it is convenient, the *summation convention* may be used from time to time.

A vector **a** is described by the components $a_x$, $a_y$, $a_z$ for example when a cartesian coordinate system is chosen. These three components can conveniently be referred to collectively as $a_i$ ($i = 1, 2, 3$ or $x, y, z$). Here the subscript i is spoken of as a *free index*.

No free index denotes a scalar, *one* free index denotes a vector, *two* free indices denote a matrix. For example, $a_{ij}$ where both $i(=1, 2, 3)$ and $j(=1, 2, 3)$ are free indices, denotes an element in a matrix.

The scalar product of two vectors **a**, **b** can be conveniently represented in index terms as

$$\mathbf{a} \, . \, \mathbf{b} = a_i \, . \, b_i = a_1 b_1 + a_2 b_2 + a_3 b_3.$$

Here the *repeated* index i can be seen to imply the summation shown, of terms of similar type added together. This is the *summation convention*, where the repeated index $a_i \cdot b_i$ implies the sum of terms obtained by letting i take on the values 1, 2, 3 in turn.

*Exercise:* Two-by-two matrices A, B can be denoted by $a_{ij}$, $b_{mn}$ where i, j; m, n can take the values 1, 2. Show that if C = AB then in index form $C_{ij} = a_{im}b_{mj} = a_{i1}b_{1j} + a_{i2}b_{2j}$, where now the repeated index m ensures the summation of row × column multiplication and summation.

Also, when convenient the following abbreviated notation will be adopted. A derivative such as dr/ds will be denoted by $(r)_{,s}$, and a second or higher derivative likewise. Such a notation is convenient typographically, but requires care by the reader in observing the placing of the comma (,). It is convenient also when used in conjunction with the summation convention. Some examples will be encountered later.

## 1.6 Elements of beam theory

Plates are in many respects two-dimensional beams. Hence here it is proposed to review beam theory and discuss the formulation and solution of beam problems.

A beam is a member in bending. This bending, produced by (vertical) loads acting on a (horizontal) member, is resisted by internal actions denoted by M and Q. Here M is the bending moment and Q is the transverse shear force.

Consider the element cut from a typical beam (Fig. 1.7). The length measured along the beam is denoted by x, and hence the element width by dx. Vertical force equilibrium, $F_R = 0$, leads to

$$Q + dQ - Q = p\,dx$$

$$\text{or} \quad \frac{dQ}{dx} = p. \tag{1.6.0}$$

When convenient this will be written $Q_{,x} = p$ where $d(\ )/dx = (\ )_{,x}$. The second equilibrium condition is the moment equilibrium ($M_R = 0$) condition. Selecting point A about which to take moments, then

$$M + dM - M - (Q + dQ)dx + p\,dx\left(\frac{dx}{2}\right) = 0.$$

Simplifying and omitting second order terms, this becomes

$$\frac{dM}{dx}(=M_{,x}) = Q. \tag{1.6.1}$$

The third equilibrium equation for force equilibrium, say horizontally, is automatically satisfied.

Then statics of the beam element is described by these two equations in two unknowns M, Q. For some purposes it is convenient to eliminate Q when

$$\frac{d^2M}{dx^2} = p. \tag{1.6.2}$$

This formulation emphasizes the differential nature of the problem, and has a parallel in plate analysis.

Beam problems are termed *statically determinate* if (1.6.2) can be solved, including the constants of integration, without knowledge of the beam material properties.

Consider a single span, simply supported beam of span L and loaded with a distributed load $p(x) = A\sin(\pi x/2L)$ where x is the length along the beam measured from one end. Calculate the distribution of moment (M) and shear force (Q) in the beam.

Then

$$M_{,xx} = A\sin\frac{\pi x}{2L}.$$

Note that simple support means $M = 0$ at the supports (the ends in this case). These are the *boundary conditions* for this example.

Now, integrating

$$M(x) = -A\left(\frac{2L}{\pi}\right)^2 \sin\frac{\pi x}{2L} + Bx + C.$$

The simple support conditions require that $M(0) = 0 = M(L)$, and these require $C = B = 0$.

Hence

$$M(x) = -A\left(\frac{2L}{\pi}\right)^2 . \sin\frac{\pi x}{2L}$$

and

$$Q = \frac{dM(x)}{dx}$$

$$= A\left(\frac{2L}{\pi}\right)\cos\frac{\pi x}{2L}.$$

This is an example of a statically determinate beam.

Suppose now that information about the deformation of the beam is required. To supply this, the material properties of the beam must be known. There are two (conventional) responses studied. Firstly, if the beam remains elastic then deformation is proportional to load and the bending moment (M) is proportional to curvature ($\kappa$). This latter quantity will be studied fully in the appendix. For the present it can be thought of as the inverse of the local radius into the beam as bent by the action of a constant bending moment.

Hence

$$M(x) = EI\,.\,\kappa(x) \tag{1.6.3}$$

where the constant of proportionality (EI) is dependent upon the size and shape of the cross-section, and the material type. It will be proved later that if the beam *deflects* transversely an amount w(x), then

$$\kappa(x) = \frac{d^2w}{dx^2}. \tag{1.6.4}$$

Hence

$$M(x) = EI\frac{d^2w(x)}{dx^2}. \tag{1.6.5}$$

If M is eliminated between (1.6.2) and (1.6.5) then

$$EI\frac{d^4w(x)}{dx^4} = p(x). \tag{1.6.6}$$

This equation is the standard equation of a uniform elastic beam, and has a direct analogue in elastic plate theory. By additional integrations, the value of w(x) for the sinusoidally loaded beam can be seen to be

$$w(x) = \frac{A}{EI}\left(\frac{2L}{\pi}\right)^4 \sin\frac{\pi x}{2L}$$

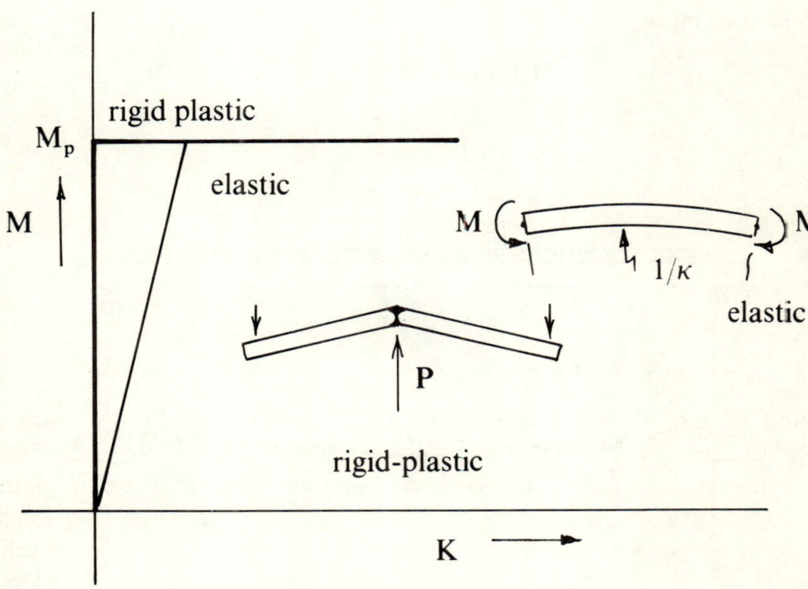


**Figure 1.8** Moment-curvature relations.

where the additional boundary conditions $w(0) = 0 = w(L)$ have been incorporated.

A second type of material response is also of interest—the plastic response when the beam is subjected to maximum load. Now the beam element may sustain a maximum moment $M = M_p$, the so-called "full plastic moment", but no more. Locally the beam member bends where $M = M_p$, the so-called "plastic hinge". When this occurs the transverse displacements become much larger than those associated with the elastic response—the beam is thought of as deflecting plastically by hinge formation, and prior elastic displacements are small and are usually neglected at this stage. This is the so-called "rigid-plastic" response. In curvature terms, the elastic and rigid plastic responses can be expressed as shown in Fig. 1.8.

Earlier, an example of an elastic beam with a sinusoidal load was considered. This was solved by integration. The same techniques can be used for rigid-plastic beams, though another method is often preferred. This preferred method of solution is by use of a *work equation.*

Consider the same beam as earlier, but now loaded with a uniformly distributed load, $p = \text{const.}$ (Fig. 1.9). This beam will collapse when the maximum bending moment reaches the full plastic value. Now by integration of (1.6.2), the maximum M is at $x = L/2$ with a value of

$$M(L/2) = -\frac{pL^2}{8}.$$

Hence at collapse

$$M(L/2) = -M_p = -\frac{pL^2}{8}$$

or

$$p_c = \frac{8M_p}{L^2},$$

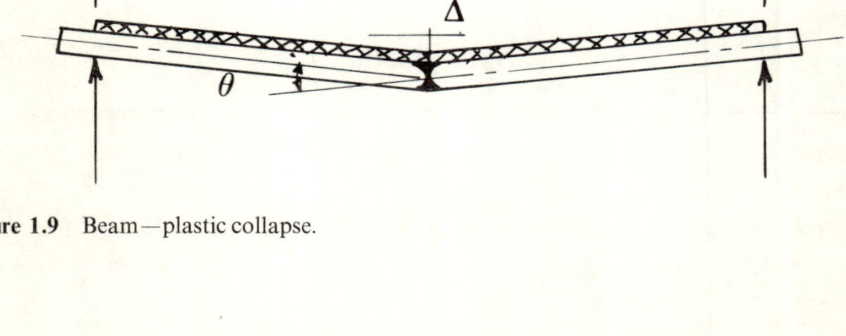


**Figure 1.9** Beam—plastic collapse.

where positive moment is assumed to be a hogging moment. This result can be obtained from first principles by allowing the collapse pressure, $p_c$, to do work on the collapse mode, with a maximum displacement of $\Delta$, this external work being absorbed as internal work by the internal plastic moment ($M_p$) working through the hinge rotation, $\theta$.

Expressed in symbols this becomes

$$2p_c \int_0^{L/2} \Delta\left(\frac{2x}{L}\right).\,dx = M_p\left(\frac{4\Delta}{L}\right)$$

or

$$4\left(\frac{p_c}{L}\right)\frac{1}{2}\left(\frac{L^2}{4}\right)\Delta = M_p\left(\frac{4\Delta}{L}\right)$$

finally

$$p_c = \frac{8M_p}{L^2}.$$

This type of work calculation is particularly useful in plate applications where the equivalent process to integration can be difficult, or even impossible.

*Exercise:* (1) Show that for the simple beam with a uniform, distributed load, p, the deflection at the beam centre will be

$$w(L/2) = \frac{5}{384}\frac{pL^4}{EI}$$

for an elastic response to load. (Hint: integration of (1.6.6) is the preferred method).

(2) For a simple beam, span L, but now restrained at the ends against rotation, show that when a concentrated load of W is applied at mid-span and increased to the maximum collapse value, $W_c$, then

$$W_c = \frac{8M_p}{L} \quad \text{and} \quad \delta_c = \frac{M_pL^2}{24EI}$$

($\delta_c$ = central deflection as collapse starts, assuming elastic/plastic response.)

In this case all the plastic hinges form simultaneously. If the load, W, is non-central, at distances a, b from the ends, $a+b = L$, $a > b$, show that $W_c.a.b = 2M_p.L$ and $6EI.\delta_c = M_p.b^2$.

## 1.7 Conclusions

In this chapter a number of topics have been introduced which are essentially preliminary to a discussion of plate theory proper. Not all the topics need be mastered initially—for example the vector manipulations. However, later, when these topics can be used with some facility, they will allow the reader a fuller exploration of the whole subject of plate theories.

# 2 Statics and kinematics of plate bending

## 2.0 Introduction

In the Appendix the geometry of surfaces is developed. Relevant results will be used here for purposes of developing a suitable plate theory. The themes are the *statics* of a bent plate and the *kinematics* of plate bending. As remarked earlier, plates are essentially two-dimensional beam-type structures. In particular, the main interest is in plates which span transversely (horizontally say) and are loaded laterally (vertically say). Such a configuration produces a *bending* system in the plate.

For the present, all plates considered will be assumed to be homogeneous. Then the middle plane, that plane which is equidistant from the two outer faces of the plate, will remain *unstretched* by the bending and twisting actions. Parts of the outer layers will be compressed, extended (and sheared by twisting actions), but though the middle surface will bend into some surface shape it will not be stretched or sheared. The theories developed are all *small deflection theories*. This is equivalent to the geometrical requirement that the surface slopes are always small, that is, much less than unity.

## 2.1 The stress resultants

The plates of interest are those subjected to bending arising from transverse loading on the plate. There is a gradient of stress through the plate, but for purposes of formulating a working theory, stresses as such will not be used. Instead, certain important *stress resultants* will be defined. The same situation exists in beam theory where shear forces and moments are used. On a cut face in the plate material, $x = \text{const.}$ (Fig. 2.0) let the stresses be referred to a right-handed cartesian system of axes $(x, y, z)$. Then $\sigma_{xy}$ is a shear stress, and acts on a face $x = \text{const.}$ and in the y-direction, whereas $\sigma_{xx}$ is a direct tension or compression (see Fig. 2.0). Since the section is being bent, $\sigma_{xx}$ is likely to be compressive on one side of the mid-plane and tensile on the other. The same may also be true of the shear components. "Stress resultant" is the general

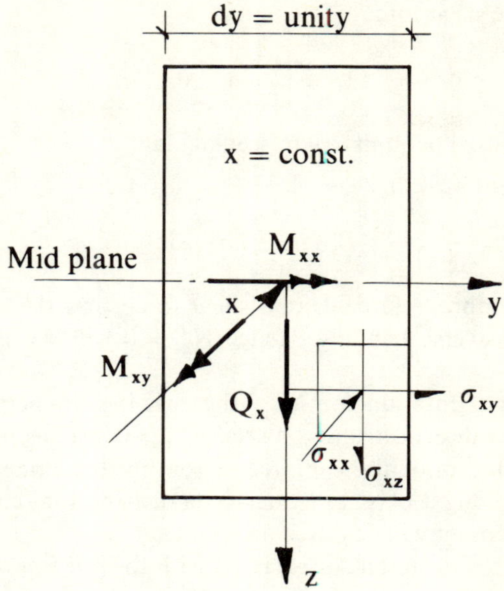


**Figure 2.0** Stress and stress resultants, $x$ = const.

term for some integral relation of stress through the thickness. Thus a typical stress resultant is

$$N_{xx} \equiv \int_{-t/2}^{t/2} \sigma_{xx} \, . \, dz. \tag{2.1.0}$$

This is the overall compressive force on the segment shown and in the present case, this force will be zero, because the plate is being bent rather than compressed. However, the quantity

$$M_{xx} \equiv \int_{-t/2}^{t/2} \sigma_{xx} z \, . \, dz \tag{2.1.1}$$

is in general not zero, and is a quantity of primary interest. This is the bending moment per unit length seen in plan and can be represented by the (double-headed) moment vector of turning *about* an axis parallel to the y axis (see Fig. 2.0).

The remaining quantities of interest are the twisting moment

$$M_{xy} \equiv \int_{-t/2}^{t/2} \sigma_{xy} z \, . \, dz \tag{2.1.2}$$

and the transverse shear force

$$Q_x \equiv \int_{-t/2}^{t/2} \sigma_{xz} \, . \, dz. \tag{2.1.3}$$

Each is also measured per unit length seen in plan.

The quantity

$$N_{xy} \equiv \int_{-t/2}^{t/2} \sigma_{xy} \, . \, dz \tag{2.1.4}$$

is, like $N_{xx}$, a membrane force, here a shear force, and these forces are not generated by transverse bending. Hence $N_{xy} = 0$ will be the situation for plates in bending.

The vector representation for $M_{xy}$ (the twisting moment) is a turning action about the x direction across a section x = const., as indicated in Fig. 2.0, and $Q_x$ is a conventional shear force vector directed along the z direction. The $M_{xy}$ can be thought of as generated by a distribution of Q-type force of amount $M_{xy}$; this point will be raised again later.

Thus far the stress resultants associated with the cut face x = const. have

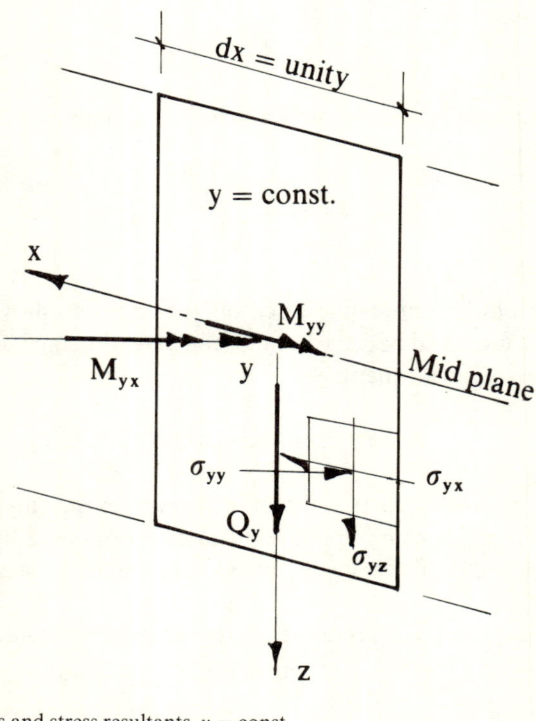


**Figure 2.1** Stress and stress resultants, $y$ = const.

been illustrated. Assuming still a right-handed cartesian coordinate system, then a cut on face y = const. must also be considered, when it will be seen (Fig. 2.1) that the relevant stress resultants are $M_{yy}$, $M_{yx}$ and $Q_y$. By comparison with the quantities on the face x = const., there follow the definitions

$$
\begin{aligned}
M_{yy} &\equiv \int_{-t/2}^{t/2} \sigma_{yy} z \, . \, dz, \\
M_{yx} &\equiv \int_{-t/2}^{t/2} \sigma_{yx} z \, . \, dz, \\
Q_y &\equiv \int_{-t/2}^{t/2} \sigma_{yz} \, . \, dz.
\end{aligned}
\tag{2.1.5}
$$

Note however that $\sigma_{xy} = \sigma_{yx}$ is an equilibrium requirement for the small element of material and hence

$$M_{xy} = M_{yx}.$$

To summarize then, the state of stress at a point in a plate will be represented by a group of stress resultants, the bending moments $M_{xx}$, $M_{yy}$; the twisting moments $M_{xy} = M_{yx}$ and the transverse shear forces $Q_x$, $Q_y$. All these quantities are measured per unit length of the mid-surface. Hence, seen in vector terms, the cut faces of a piece of plate will be subjected to internal forces and moments as indicated in Fig. 2.2. The implied sign convention is then that *hogging* bending moments are *positive.*

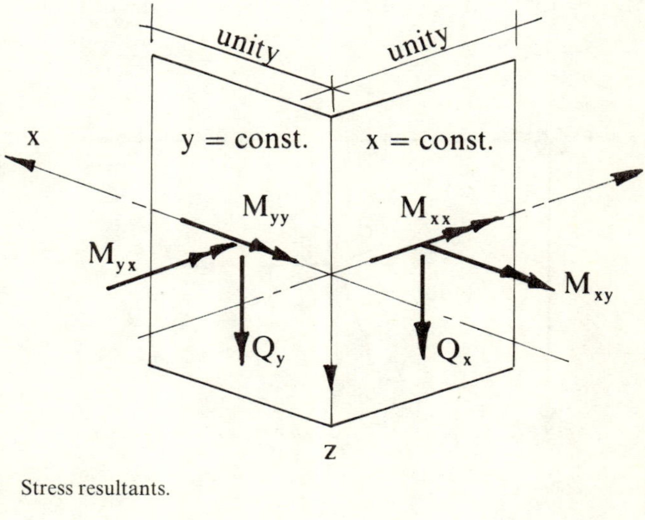


**Figure 2.2** Stress resultants.

The collection of moments $M_{xx}$, $M_{xy} = M_{yx}$, $M_{yy}$ can conveniently be written as a $2 \times 2$ matrix, the moment matrix, thus

$$\mathbf{M} \equiv \begin{bmatrix} M_{xx} & M_{xy} \\ M_{xy} & M_{yy} \end{bmatrix}. \tag{2.1.6}$$

This is a symmetric matrix and once the three independent components of this matrix are known at a point for one choice of axes, then the values of the bending and twisting moment on any other transverse plane through this point, different from the x = const. and y = const. planes, can be found. Just how this can be achieved will now be considered.

## 2.2 Principal values

The moment matrix, **M**, when referred to x and y axes at a point P, is written

$$\mathbf{M} = \begin{bmatrix} M_{xx} & M_{xy} \\ M_{xy} & M_{yy} \end{bmatrix}. \tag{2.2.0}$$

Suppose it is desired to refer the matrix **M** to the choice of axes x′y′ at P (Fig. 2.3). What can be said about the values of the elements in the corresponding matrix **M**′? One way of proceeding is to consider the equilibrium of a small triangular element notionally cut from the plate and containing P (Fig. 2.4*a*,*b*).

Recall first that all the quantities are computed per unit length, then resolve along, and normal to, x′ when

$$\begin{aligned} M_{y'y'} &= M_{yy}\,.\operatorname{Cos}^2\theta + M_{xx}\operatorname{Sin}^2\theta - 2M_{xy}\operatorname{Sin}\theta\operatorname{Cos}\theta \\ M_{x'y'} &= (M_{yy} - M_{xx})\operatorname{Sin}\theta\operatorname{Cos}\theta + M_{xy}(\operatorname{Cos}^2\theta - \operatorname{Sin}^2\theta). \end{aligned} \tag{2.2.1}$$

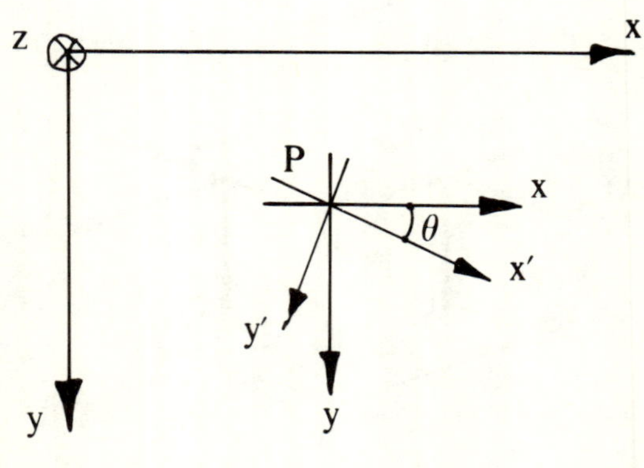


**Figure 2.3** Rotation of axes.

Again, resolve along and normal to y′ direction (Fig. 2.4*b*), then

$$M_{x'x'} = M_{xx}\,\mathrm{Cos}^2\,\theta + M_{xy}\,\mathrm{Sin}^2\,\theta + 2M_{xy}\,\mathrm{Sin}\,\theta\,\mathrm{Cos}\,\theta$$
$$M_{x'y'} = (M_{yy} - M_{xx})\,\mathrm{Sin}\,\theta\,\mathrm{Cos}\,\theta + M_{xy}(\mathrm{Cos}^2\,\theta - \mathrm{Sin}^2\,\theta). \tag{2.2.2}$$

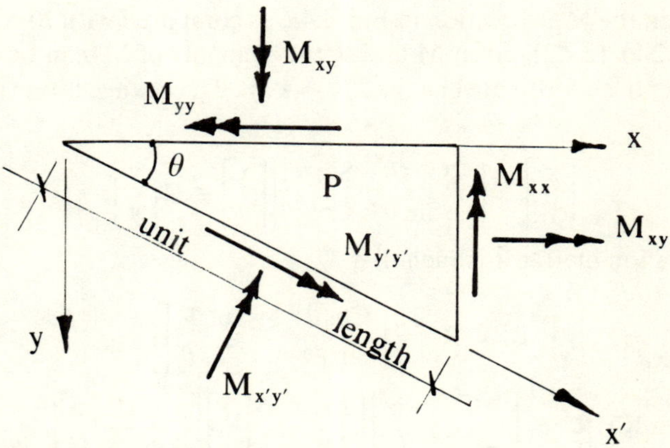


**Figure 2.4**(*a*) Change of axes I.

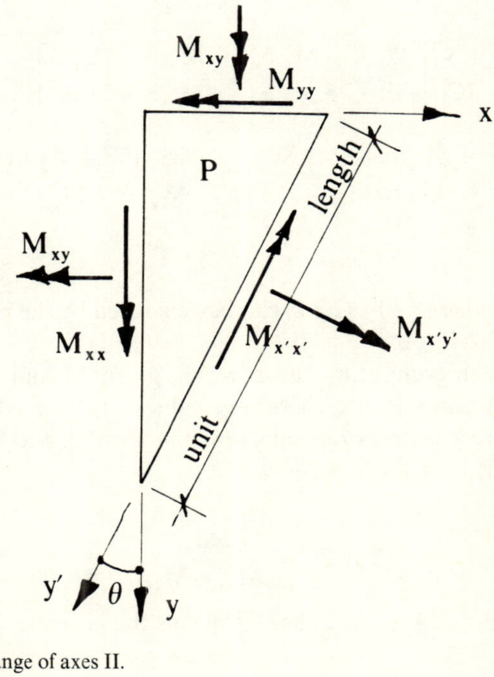


**Figure 2.4**(*b*) Change of axes II.

It is worth pausing to consider the directions shown for quantities in Fig. 2.4*a*,*b*. By construction, all the edges of the triangular elements are along coordinate lines. The basic directions of quantities have been carried forward from Fig. 2.1 and 2.2, but where the face concerned bounds the far rather than the near side of this element, then by Newton's Third Law of Action-Reaction, reversed quantities are introduced. Such considerations account for the change in the $M_{xx}$ direction in Fig. 2.4*a*, as compared with Fig. 2.2.

From (2.2.1), (2.2.2), given **M** and $\theta$, the elements of **M**′ can be evaluated. But consider the coordinate change $x, y \rightarrow x', y'$. The change is one of rotation where

$$\begin{bmatrix} x' \\ y' \end{bmatrix} = \begin{bmatrix} \text{Cos}\,\theta & \text{Sin}\,\theta \\ -\text{Sin}\,\theta & \text{Cos}\,\theta \end{bmatrix} \begin{bmatrix} x \\ y \end{bmatrix} \equiv \mathbf{R} \begin{bmatrix} x \\ y \end{bmatrix}, \qquad (2.2.3)$$

and the rotation matrix **R** is such that

$$\mathbf{R}^{I} = \mathbf{R}^{T} = \begin{bmatrix} \text{Cos}\,\theta & -\text{Sin}\,\theta \\ \text{Sin}\,\theta & \text{Cos}\,\theta \end{bmatrix}.$$

Now $$\mathbf{RMR}^{T} = \begin{bmatrix} C & +S \\ -S & C \end{bmatrix} \begin{bmatrix} M_{xx} & M_{xy} \\ M_{xy} & M_{yy} \end{bmatrix} \begin{bmatrix} C & -S \\ +S & C \end{bmatrix},$$

where $C = \text{Cos}\,\theta$, $S = \text{Sin}\,\theta$.

Thus

$$\mathbf{RMR}^{T} = \begin{bmatrix} C & +S \\ -S & C \end{bmatrix} \begin{bmatrix} M_{xx}C + M_{xy}S & -M_{xx}S + M_{xy}C \\ M_{xy}C + M_{yy}S & -M_{xy}S + M_{yy}C \end{bmatrix}$$

$$= \begin{bmatrix} M_{xx}C^2 + M_{yy}S^2 + 2M_{xy}SC & M_{xy}(C^2 - S^2) + (M_{yy} - M_{xx})SC \\ M_{xy}(C^2 - S^2) + (M_{yy} - M_{xx})SC & M_{xx}S^2 + M_{yy}C^2 - 2M_{xy}SC \end{bmatrix}$$

$$= \mathbf{M}' \qquad (2.2.4)$$

Hence the formulae (2.2.1), (2.2.2) can be generated by the matrix similarity transformation (2.2.4).

When engaged in computing the new **M**′, given **M** and $\theta$, the question which (naturally) arises is, are there any values of $\theta$ for which $M_{x'y'} \equiv 0$? Such directions are known as *principal directions* for **M**, and from (2.2.1, 2) it can be seen that $M_{x'y'} \equiv 0$ when $\theta = \alpha$.

Then

$$\text{Tan}\,2\alpha = \frac{2M_{xy}}{(M_{xx} - M_{yy})}. \qquad (2.2.5)$$

The corresponding $M_{x'x'}$, $M_{y'y'}$ are known as the *principal bending moment values.*

*Exercises* (1) Note the similarities between $M_{\alpha\beta}$ moment fields and fields at stress and strain in two dimensions.

(2) Suppose in x, y coordinates that at a point in a plate there is pure twist, $M_{xx} = M_{yy} = 0$, $M_{xy} = M_{yx} = m$. Find the principal bending moments, and the planes on which they act.

Now

$$\mathbf{M} = \begin{bmatrix} 0 & m \\ m & 0 \end{bmatrix}.$$

It is required to find a new $\mathbf{M}'$ such that

$$\mathbf{M}' = \begin{bmatrix} \mu_1 & 0 \\ 0 & \mu_2 \end{bmatrix},$$

where $\mu_{1,2}$ are the principal moments. Hint: The appropriate x′, y′ are inclined at $\pi/4$ to x, y. *Answer:* $\mu_1 = m = -\mu_2$.

## 2.3 The moment circle

The algebraic manipulations of the previous section can alternatively be carried through to the same result by means of a geometrical construction, the moment circle construction (Fig. 2.5).

The axes for the circle are bending moment (horizontal) and twisting moment (vertical). Consider the planes associated with the x, y axis system, Fig. 2.2. Then the bending moment $M_{xx}$ and twisting moment $M_{xy}$ act on the plane x = const., whereas the bending moment $M_{yy}$ and twisting moment $M_{yx}(=M_{xy})$ act on the plane y = const. Treated as coordinate pairs, two points can now be plotted on the B.M./T.M. axes of the *moment circle*.

One of the $M_{xy}$'s is given a negative sign—suppose it is the one associated with $M_{yy}$. Then points x, y, as shown on Fig. 2.5, can be plotted. The mid-point of the line joining x and y is labelled P, and a circle centre P, radius PX(=PY) is drawn. This circle is the moment circle.

By simple geometry on Fig. 2.5 the expression for $\theta$ to give principal bending moments can be seen to follow (2.2.5) thus giving the points marked I and II on Fig. 2.5 as the principal bending moments. Note that these bending moments are associated with zero twisting moment, by definition, and that they are algebraically the greatest and least bending moments acting at the point P, selected from amongst all the possible planes drawn through the plate thickness.

Note too that the construction carries data for the *planes* along x and y directions into *points* x and y on the circle which are on the ends of a diameter. Now the planes x and y subtend an angle of $\pi/2$ at P, where the points x and y subtend an angle double this, $\pi$, at the circle centre. Likewise

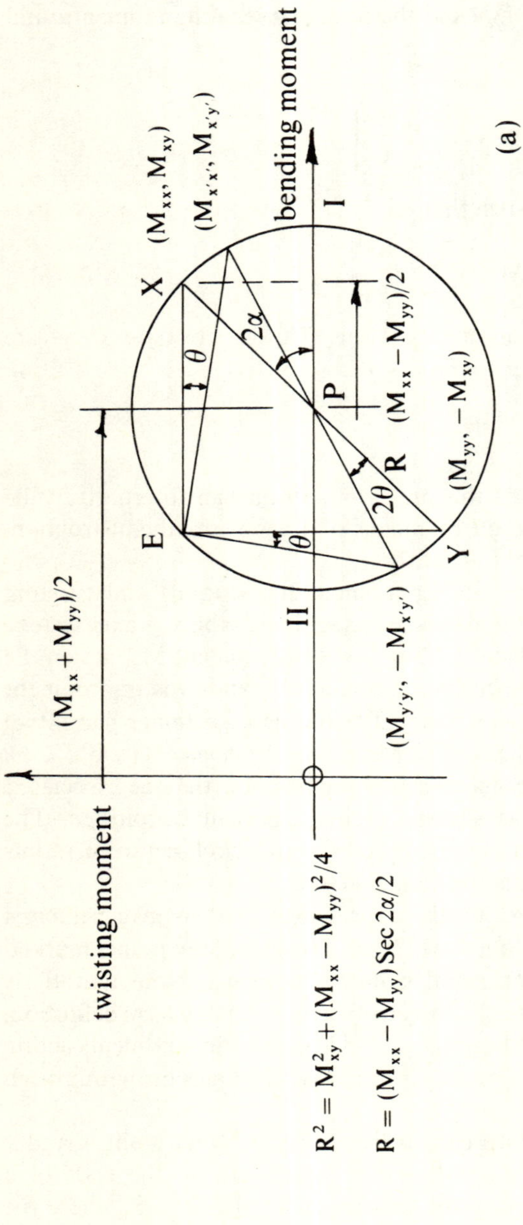


**Figure 2.5**(*a*) Moment circle.

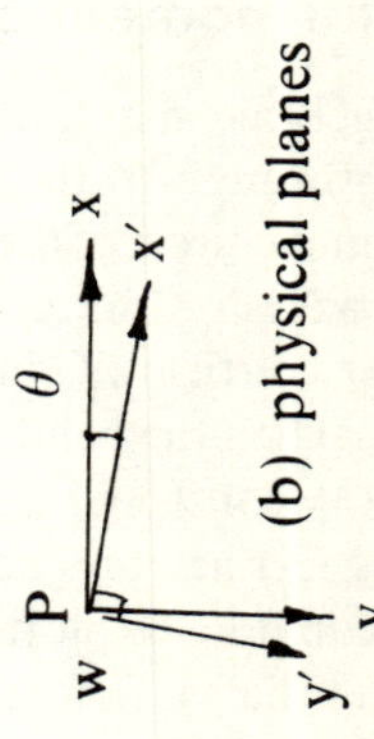


**Figure 2.5**(*b*) Physical axes.

the *principal moments*, represented by I, II on the circle act on planes through P which are normal to one another, since these planes must subtend at P, half the angle which I and II subtend at the circle centre.

A useful additional feature of the moment circle is the pole point, E, on the perimeter. This is constructed by drawing, through the points x and y respectively, lines which are parallel to the *physical* planes on which the respective moment/twist combinations act. Shown in Fig. 2.5*b* are the physical plane directions—x, y; x′, y′. Now by construction, since the x, y planes are normal, the additional lines now drawn are also normal to one another, and this means that they must intersect *on* the moment circle in order that they may enclose this angle. Then, knowing E, the joins to X′ and Y′ give directions which are parallel to the physical planes on which the $M'_{xx}$, $M'_{xy}$ etc. act— see also Fig. 2.5*b*.

## 2.4 Equilibrium equations—rectangular coordinates

Thus far the bending and twisting moments at a point in the plate have been considered, and the question of change of reference axes at the point has been examined. Now an element of the plate is to be considered, and the requirements that this element be in equilibrium will be established. For this a coordinate system must be chosen. Here the simplest system, the rectangular cartesian system, will be adopted. Later another possible choice will be investigated.

Consider the plate element, side lengths dx, dy, as shown in Fig. 2.6. If the plate element is to support any loading, such as a pressure, p(x, y), then not all the internal forces can be constant.

Typical variation of the internal stress resultants is shown in Fig. 2.6. Each of $M_{xx}, M_{yy} : M_{xy}, M_{yx}$; $Q_x, Q_y$ is shown as varying between the edges of the element, in the manner of the differential calculus. Now apply the equations of statics to this element to determine the equilibrium equations.

Recall first that all the stress resultants are defined per unit length along the middle surface (as seen in plan). Next note that *three* independent equations of statics can be written by choosing *two* distinct axes in the x, y plane about which to take moments, together with a third equation describing the *force* equilibrium in the z direction.

First choose the axis of y about which to take moment equilibrium. Then

$$-M_{xx}\,.\,dy+\left(M_{xx}+\frac{\partial M_{xx}}{\partial x}dx\right)dy-M_{yx}\,dx+\left(M_{yx}+\frac{\partial M_{yx}}{\partial y}dy\right)dx$$
$$-\left(Q_x+\frac{\partial Q_x}{\partial x}dx\right)dy\,.\,dx+p\,.\,(dx)^2\,.\,\frac{dx}{2}=0.$$

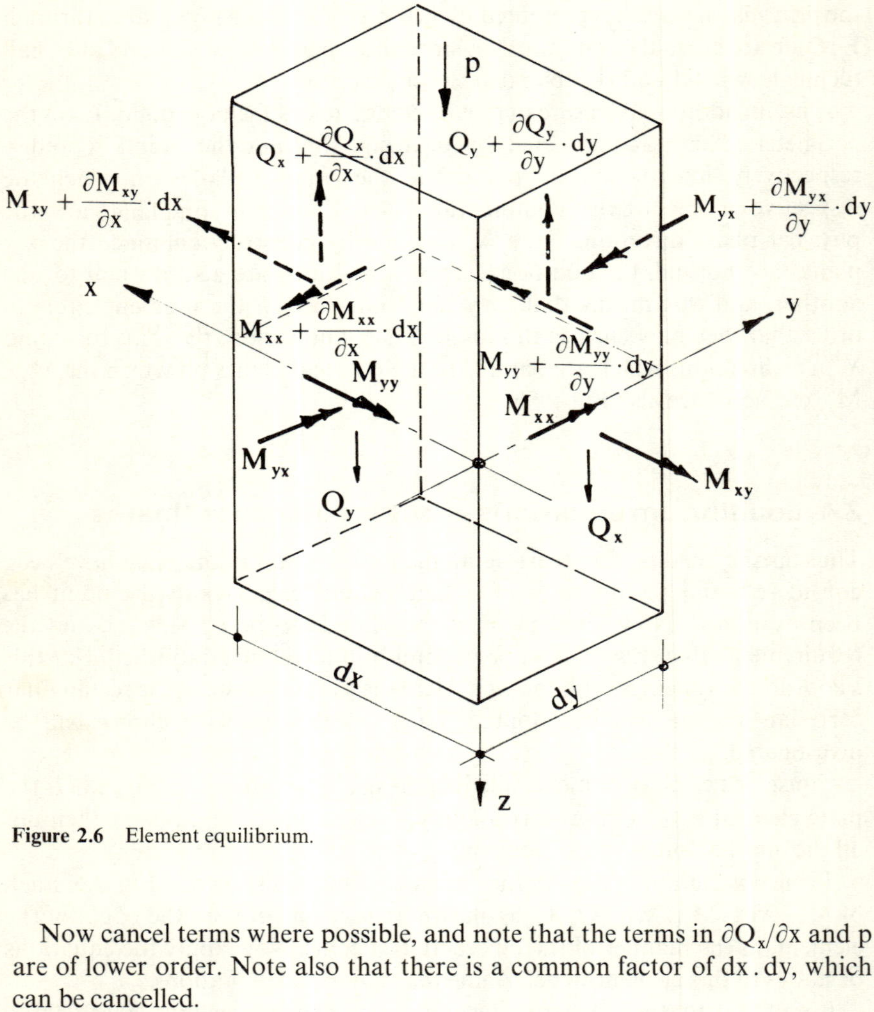


**Figure 2.6** Element equilibrium.

Now cancel terms where possible, and note that the terms in $\partial Q_x/\partial x$ and p are of lower order. Note also that there is a common factor of dx . dy, which can be cancelled.

Then the resulting equation of moment equilibrium is

$$\frac{\partial M_{xx}}{\partial x} + \frac{\partial M_{yx}}{\partial y} - Q_x = 0. \tag{2.4.0}$$

By a similar argument, but with respect to the x axis, there is obtained a second moment equilibrium equation

$$\frac{\partial M_{xy}}{\partial x} + \frac{\partial M_{yy}}{\partial y} - Q_y = 0. \tag{2.4.1}$$

The third and final equilibrium equation is for forces in the z direction.

Consider, then, the whole element. The forces contributing are the various $Q_x$, $Q_y$ forces and the applied pressure loading p(x, y).

Thus

$$\left[\left(Q_x+\frac{\partial Q_x}{\partial x}dx\right)-Q_x\right]dy+\left[\left(Q_y+\frac{\partial Q_y}{\partial y}dy\right)-Q_y\right]dx = p\,dx\,.\,dy.$$

Simplifying, and cancelling by dx . dy,

$$\frac{\partial Q_x}{\partial x}+\frac{\partial Q_y}{\partial y}=p. \tag{2.4.2}$$

These three equations involve five unknowns, $M_{xx}$, $M_{xy} = M_{yx}$, $M_{yy}$; $Q_x$ and $Q_y$. Hence the plate moments are *indeterminate*, in general.

For some purposes it is useful to eliminate the transverse shear forces $Q_x$, $Q_y$. When this is done a single equation in the three bending stress resultants is obtained. To obtain this, differentiate (2.4.0) with respect to x, (2.4.1) with respect to y and eliminate $Q_x$, $Q_y$ between these two and (2.4.2) to give

$$\frac{\partial^2 M_{xx}}{\partial x^2}+2\frac{\partial^2 M_{xy}}{\partial x\,\partial y}+\frac{\partial^2 M_{yy}}{\partial y^2}=p. \tag{2.4.3}$$

Note the similarity between this equation and the beam equilibrium equation (1.6.2).

*Worked example:* Show that the field of moments

$$M_{xx} = -\frac{pL^2}{8}\left(1-\left(\frac{x}{L}\right)^2\right)$$

$$M_{xy} = \frac{p}{4}xy$$

$$M_{yy} = -\frac{pL^2}{8}\left(1-\left(\frac{y}{L}\right)^2\right)$$

is in equilibrium with a pressure loading, p = const., and investigate the principal moments at (O, O), (O, L) and (L, L).

Now

$$\frac{\partial^2 M_{xx}}{\partial x^2}=\frac{p}{4},\frac{\partial^2 M_{xy}}{\partial x\,\partial y}=\frac{p}{4},\frac{\partial^2 M_{yy}}{\partial y^2}=\frac{p}{4}.$$

Hence equation (2.4.3) is satisfied.

At (O, O) $M_{xx} = M_{yy} = -pL^2/8$, $M_{xy} = 0$. Hence $M_{xx}$, $M_{yy}$ are principal at the origin. At (O, L) $M_{xx} = 0$, $M_{xy} = 0$, $M_{yy} = -pL^2/8$. Hence again $M_{xx}$ and $M_{yy}$ are principal.

At (L, L) $M_{xx} = M_{yy} = 0$ and $M_{xy} = pL^2/4$.

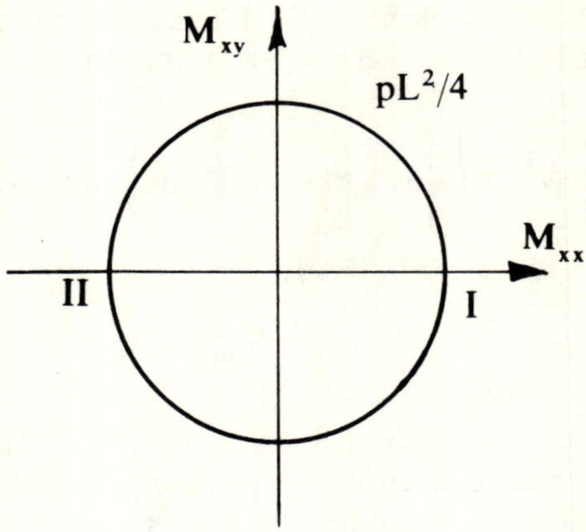


**Figure 2.7** Example—moment circle.

Thus the principal axes are at $\pi/4$ to x, y and the principal moments are $\pm pL^2/4$. (See Fig. 2.7).

## 2.5 Plate bending kinematics—rectangular coordinates

The term kinematics is used to describe the deformations of the plate. When loaded, the plate deflects, carrying the original plane middle surface into some shallow curved shape. This surface will in general have no special features.

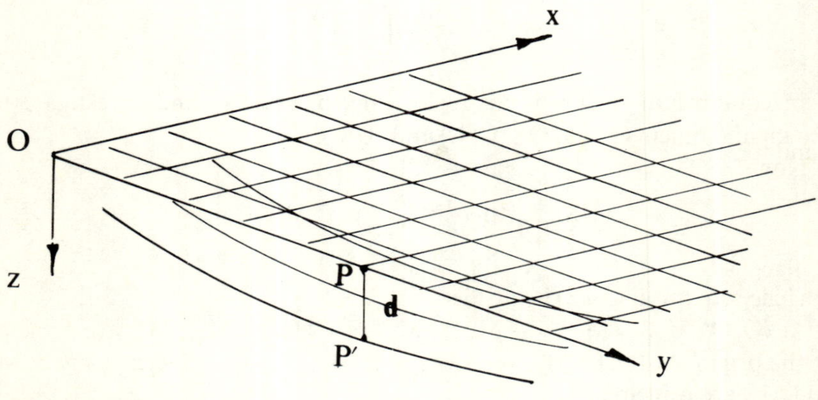


**Figure 2.8** Plate deformation—displacement.

There will be slopes and curvatures at every point and in every direction through every point on the deflected surface. But the deflected shape at the plate will be *shallow* in the sense that everywhere the slopes will be small compared to unity.

As a result, a number of approximations can be introduced to simplify the description of the curved deflected shape of the middle surface. Our purpose here is to decide what geometrical information about this shape will be needed for the construction of physically reasonable plate theories. Recall that it is intended to use a rectangular cartesian coordinate system (Fig. 2.8).

Suppose a typical point (P) on the plate mid surface displaces to P′ as the plate deforms under load. This displacement will be denoted by w(x, y). As a consequence of the shallowness of the deflected surface, the typical point P′ will have no components of displacement in either the x or y directions.

The vector displacement of P to P′ is therefore

$$\mathbf{d} = \mathrm{w}\mathbf{k}. \tag{2.5.0}$$

Hence, with respect to an x, y, z coordinate system the deformed middle surface of the plate is described by a radius vector $\mathbf{r}_P$ given by

$$\mathbf{r}_P = \mathrm{x}\mathbf{i} + \mathrm{y}\mathbf{j} + \mathrm{w}\mathbf{k}. \tag{2.5.1}$$

Here x and y are to be regarded as the independent variables—the surface coordinates.

Knowing the form of the radius vector to a typical point P, the procedures discussed in the Appendix can now be used to evaluate all the relevant geometrical quantities.

Thus the tangent vectors, $\mathbf{r}_x$, $\mathbf{r}_y$, (see Appendix, section A.2.2) are given by

$$\mathbf{r}_x \equiv \frac{\partial \mathbf{r}}{\partial x} = \mathbf{i} + \frac{\partial w}{\partial x}.\mathbf{k},$$

$$\mathbf{r}_y \equiv \frac{\partial \mathbf{r}}{\partial y} = \mathbf{j} + \frac{\partial w}{\partial y}.\mathbf{k},$$

and the coefficients of the first fundamental form, E, F, G, (A.2.4) are given by

$$\begin{aligned} E &\equiv \mathbf{r}_x.\mathbf{r}_x = 1 + \left(\frac{\partial w}{\partial x}\right)^2, \\ F &\equiv \mathbf{r}_x.\mathbf{r}_y = \left(\frac{\partial w}{\partial x}\cdot\frac{\partial w}{\partial y}\right), \\ G &\equiv \mathbf{r}_y.\mathbf{r}_y = 1 + \left(\frac{\partial w}{\partial y}\right)^2. \end{aligned} \tag{2.5.2}$$

Recall now the notion of shallowness; this implies that

$$\left(\frac{\partial w}{\partial x}\right), \left(\frac{\partial w}{\partial y}\right) \ll 1 \quad \text{then}$$

$$\begin{aligned} E &\doteqdot 1, \\ F &\doteqdot 0, \\ G &\doteqdot 1. \end{aligned} \tag{2.5.3}$$

These approximate expressions indicate that the original coordinate lines on the deformed plate middle surface are still sensibly *orthogonal*.

Proceed to consider the unit surface normal (A.3) and the surface curvatures. Now **n**, the unit normal, is defined by (A.3.0) as

$$\mathbf{n} \equiv \frac{\mathbf{r}_x \times \mathbf{r}_y}{|\mathbf{r}_x \times \mathbf{r}_y|}.$$

In this case

$$\mathbf{r}_x \times \mathbf{r}_y = -\frac{\partial w}{\partial x}\mathbf{i} - \frac{\partial w}{\partial y}\mathbf{j} + 1\,.\,\mathbf{k},$$

and

$$|\mathbf{r}_x \times \mathbf{r}_y| = \left[1 + \left(\frac{\partial w}{\partial x}\right)^2 + \left(\frac{\partial w}{\partial y}\right)^2\right]^{1/2} \doteqdot 1.$$

Hence

$$\mathbf{n} = -\frac{\partial w}{\partial x}\mathbf{i} - \frac{\partial w}{\partial y}\mathbf{j} + \mathbf{k}. \tag{2.5.4}$$

Next the coefficients of the second fundamental form (A.4.1) are given by

$$\begin{aligned} L \equiv \mathbf{n}\,.\,\mathbf{r}_{xx} &= \left(-\frac{\partial w}{\partial x}\mathbf{i} - \frac{\partial w}{\partial y}\mathbf{j} + \mathbf{k}\right).\left(\frac{\partial^2 w}{\partial x^2}\mathbf{k}\right), \\ &= \frac{\partial^2 w}{\partial x^2} \end{aligned}$$

$$\begin{aligned} M \equiv \mathbf{n}\,.\,\mathbf{r}_{xy} &= \left(-\frac{\partial w}{\partial x}\mathbf{i} - \frac{\partial w}{\partial y}\mathbf{j} + \mathbf{k}\right).\left(\frac{\partial^2 w}{\partial x\,\partial y}\mathbf{k}\right), \\ &= \frac{\partial^2 w}{\partial x\,\partial y} \end{aligned} \tag{2.5.5}$$

and

$$\begin{aligned} N \equiv \mathbf{n}\,.\,\mathbf{r}_{yy} &= \left(-\frac{\partial w}{\partial x}\mathbf{i} - \frac{\partial w}{\partial y}\mathbf{j} + \mathbf{k}\right).\left(\frac{\partial^2 w}{\partial y^2}\mathbf{k}\right), \\ &= \frac{\partial^2 w}{\partial y^2}. \end{aligned}$$

Hence finally, recalling that the surface coordinate lines namely the x, y coordinates projected on to the deflected middle surface, are sensibly orthogonal, therefore the curvatures and twists on the deformed middle surface are given by

$$\kappa_{xx} \equiv L/E = \frac{\partial^2 w}{\partial x^2}, \qquad \text{(A.9.0)}$$

$$\kappa_{xy} \equiv \kappa_{yx} \equiv M/\sqrt{EG} = \frac{\partial^2 w}{\partial x\,\partial y}, \qquad \text{(A.9.4)} \qquad\qquad (2.5.6)$$

$$\kappa_{yy} \equiv N/G = \frac{\partial^2 w}{\partial y^2}. \qquad \text{(A.9.1)}$$

These expressions are the basic kinematic relations required for the construction of a suitable plate theory when expressed in cartesian coordinates.
*Worked example:* Consider the deflected shape

$$w(x, y) = A\,\text{Sin}\frac{\pi x}{L}\text{Sin}\frac{\pi y}{L},$$

to be defined over the square $0 \leqslant x \leqslant L, 0 \leqslant y \leqslant L$. Then $w = 0$ all round the boundary. At a typical point x, y

$$\kappa_{xx} = -\left(\frac{\pi}{L}\right)^2 w(x, y) = \kappa_{yy} \quad \text{and} \quad \kappa_{xy} = A\left(\frac{\pi}{L}\right)^2 \text{Cos}\frac{\pi x}{L}\text{Cos}\frac{\pi y}{L}.$$

Consider the direction $x = y$, that is, a diagonal of the square. At the centre $\kappa_{xy} = 0$ since $x = y = L/2$, then $\kappa_{xx} = \kappa_{yy}$ are *principal curvatures*, and are at maximum value. At $x = y = L/4$, then $\kappa_{xx} = \kappa_{yy} = -A/2(\pi/L)^2$ and $\kappa_{xy} = +A/2(\pi/L)^2$. The curvature circle has a vertical diameter through the centre, with the x, y directions at the ends. Hence the x, y directions are always at $\pi/4$ to the principal curvature directions, except when $\kappa_{xy} = 0$, when the x, y directions are principal— and so too are all other directions.
*Exercise:* If $w = Ax(a-x)y(b-y)$ as $0 \leqslant x \leqslant a$, $0 \leqslant y \leqslant b$, derive the curvature matrix $\boldsymbol{\kappa}$ (A.10.0) and explore the properties of the principal directions.

*Answer:* $\kappa_{xx} = 2Ay(b-y), \kappa_{xy} = A(a-2x)(b-2y)$.

## 2.6 Equilibrium equations—polar coordinates—radial symmetry

Earlier, equilibrium of a plate element referred to rectangular coordinates was presented. Here polar coordinates will be considered. However only a special case will be dealt with, the case of axial symmetry. This symmetry corresponds to independence for all the dependent quantities from change of

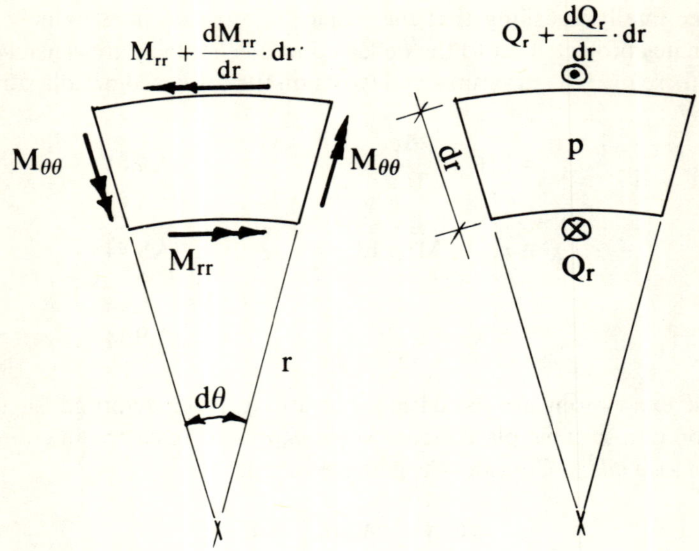


**Figure 2.9** Circular plate—element forces.

the angular coordinate, $\theta$—termed *axial symmetry*. This feature has been incorporated in the stress resultant array on Fig. 2.9.

Since only variation with respect to r is now possible, so ordinary rather than partial differentials are indicated. Also, this form of symmetry ensures that $Q_\theta = 0$. It is left as an exercise for the reader to construct and demonstrate the correctness of Fig. 2.9.

Because of the symmetry, only two independent equilibrium equations exist—one a moment equation and the other a force equation for resolution in the axial direction.

Consider moments of the forces shown in Fig. 2.9 about a tangential axis. Then

$$\left[M_{rr}+\frac{dM_{rr}}{dr}dr\right][r+dr]\,d\theta-M_{rr}\,.\,r\,d\theta-M_{\theta\theta}\,.\,2\frac{d\theta}{2}dr$$

$$-\left[Q_r+\frac{dQ_r}{dr}dr\right][r+dr]\,d\theta\,.\,dr=0.$$

Simplify and neglect the second order term in $dQ_r/dr$, then divide through by $dr\,.\,d\theta$.

Hence there is obtained

$$r\frac{dM_{rr}}{dr}+M_{rr}-M_{\theta\theta}-Q_r\,.\,r=0. \tag{2.6.0}$$

By considering force equilibrium in the z-direction, obtain

$$\left[Q_r+\frac{dQ_r}{dr}.dr\right][r+dr]\,d\theta-Q_r.r.d\theta-p.r.d\theta.dr=0,$$

or

$$r\frac{dQ_r}{dr}+Q_r=p.r. \tag{2.6.1}$$

These two equations can be rewritten as

$$\begin{aligned}(rM_{rr})_{,r}-M_{\theta\theta}&=r.Q_r,\\ (rQ_r)_{,r}&=p.r.\end{aligned} \tag{2.6.2}$$

These are two equations in three unknowns, $M_{rr}$, $M_{\theta\theta}$, $Q_r$ and hence are indeterminate.

*Exercise:* Show that $M_{rr}=p(a^2-r^2)/6$, $M_{\theta\theta}=pa^2/6$, is an equilibrium system with p = const., a = const., and find $Q_r$. *Answer:* $Q_r=pr/2$.

## 2.7 Plate bending kinematics—polar coordinates—radial symmetry

Useful theories for plate behaviour can only be constructed once both statics and kinematics are linked, because of the statical indeterminacy. In this section, the basic aim is to evaluate the deflected shape *curvatures* from a knowledge of the deflected shape itself, and in the following chapters complete theories will be discussed.

As for the rectangular coordinate description, and resulting from the shallow character of the deflected surface, the typical radius vector for a point on the plate deflected surface can now be written as

$$\mathbf{r}=r.\mathbf{I}+w(r).\mathbf{K} \tag{2.7.0}$$

where the position vector **r** is seen to depend only on r and the transverse deflection, w. The unit vectors **I** and **K** are directed radially and transversely—see Fig. 2.10.

The deflected middle surface in this case is a surface of revolution, because radial symmetry is assumed. Hence the results of Appendix A.13 are relevant. However, since the surface here is a *shallow* shape, the curvature expressions can conveniently be evaluated from first principles. Now the surface slope in the radial direction at P is given by $w_{,r}$. From symmetry, the circumferential slope is zero.

It is reasonable to suppose that the section (Fig. 2.10*a*) cuts the deflected surface along a principal curvature direction. This is confirmed by the

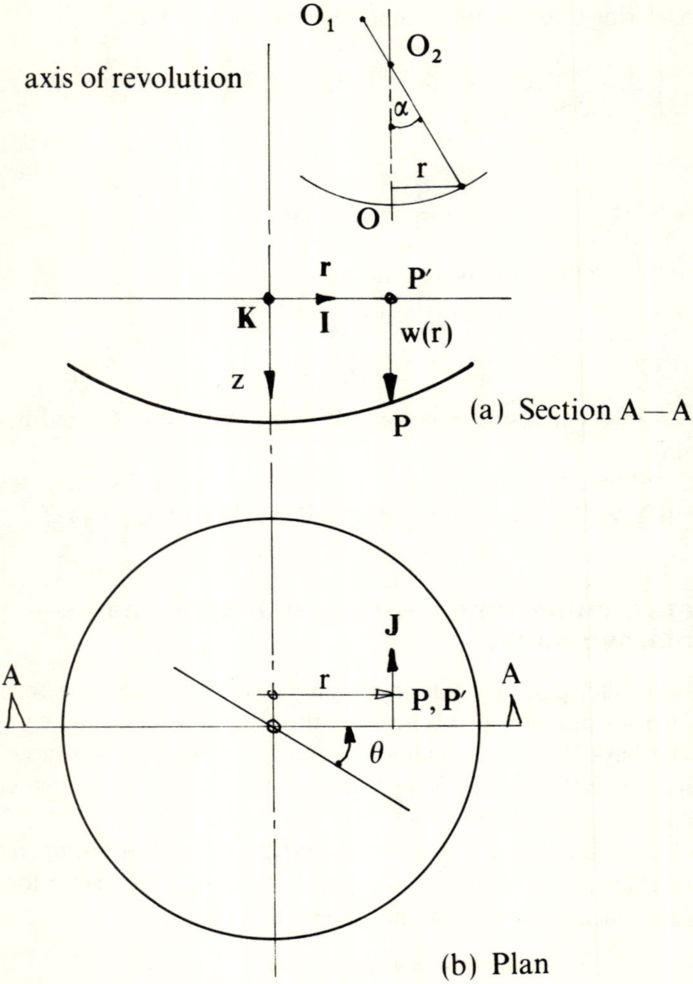


**Figure 2.10** Circular plate—deformation geometry.

analysis of section 1.13. The curvature in this plane, $\kappa_1$, is associated with a centre of curvature $O_1$, and a radius $O_1P$ (see inset, Fig. 2.10*a*). The second principal curvature is then associated with the section of the surface by a plane containing the normal at P, and the normal to the plane of the section A–A. As shown in Section 1.13 this curvature is associated with a centre of curvature $O_2$ which is given by the point where the normal intersects the axis of revolution.

Take the surface coordinates to be r and $\theta$. By this is meant that the point P is located in plan by the choice at r and $\theta$, then P′, the point required on the deflected surface is vertically below P, a distance w(r), the amount by which the point originally at P deflects to P′.

Now, the operation $(..)_{,r}$ does not affect any of the orthogonal unit vectors **I**, **J**, **K**; however, $(..)_{,\theta}$ does.

Thus

$$\mathbf{I}_{,\theta} = \mathbf{J}, \quad \mathbf{J}_{,\theta} = -\mathbf{I}, \quad \mathbf{K}_{,\theta} = 0. \tag{2.7.1}$$

Hence,

$$\mathbf{r}_{,r} = \mathbf{I} + w_{,r}\mathbf{K},$$

$$\mathbf{r}_{,\theta} = r\mathbf{J} + \mathrm{O}.$$

The coefficients of the first fundamental form (A.2.4) are then

$$E = \mathbf{r}_{,r} . \mathbf{r}_{,r} = 1 + w_{,r}^2 \doteqdot 1$$

$$F = \mathbf{r}_{,r} . \mathbf{r}_{,\theta} = 0,$$

$$G = r^2.$$

Now

$$\mathbf{n} = \frac{\mathbf{r}_{,r} \times \mathbf{r}_{,\theta}}{|\mathbf{r}_{,r} \times \mathbf{r}_{,\theta}|} = \frac{-r . w_{,r} . \mathbf{I} + r . \mathbf{K}}{(r^2 + r^2 w_{,r}^2)^{1/2}}$$

but $w_{,r} \ll 1$, hence

$$\mathbf{n} = -w_{,r} . \mathbf{I} + \mathbf{K}.$$

Then the second fundamental form coefficients (A.4.0, A.4.1) are

$$L \equiv \mathbf{n} . \mathbf{r}_{,rr} = \mathbf{n} . (w_{,rr})\mathbf{K},$$

$$= w_{,rr};$$

$$M \equiv \mathbf{n} . \mathbf{r}_{,r\theta} = \mathbf{n} . \mathbf{J} = 0;$$

$$N \equiv \mathbf{n} . \mathbf{r}_{,\theta\theta} = \mathbf{n} . -r\mathbf{I} = r w_{,r}.$$

The $r, \theta$ directions are therefore principal, because for this choice of coordinates $F = M = 0$.

Hence the principal curvatures (A.10.1) are given by

$$\kappa_1 = L/E = w_{,rr},$$

$$\kappa_2 = N/G = \frac{w_{,r}}{r}. \tag{2.7.2}$$

The curvature, $\kappa_1$, is that associated with the radial section, Fig. 2.10*a*, and the curvature centre $O_1$, while the curvature $\kappa_2$ is the curvature of a section transverse to the previous one and is associated with the curvature $O_2$.

These curvature values can be confirmed by direct calculation using the properties of $O_1$ and $O_2$.

Thus from Fig. 2.10 (inset), since the curve OP is shallow, the curvature of OP at P is given by $w_{,rr}$ as for a beam. Also from the figure, $O_2P = r/\text{Sin}\,\alpha$ and $\alpha$ is small, being such that $\text{Tan}\,\alpha = w_{,r} \ll 1$. Hence $\text{Sin}\,\alpha = \text{Tan}\,\alpha = w_{,r}$, and $O_2P = r/w_{,r}$, or the curvature

$$\kappa_2 = \frac{1}{O_2P} = w_{,r}/r.$$

*Exercise:* Suppose $w = C(a^2 - r^2)^2$, $a = \text{const.}$, show that the principal curvatures are given by

$$\kappa_{rr} = 4C\,.\,(3r^2 - a^2), \quad \kappa_{\theta\theta} = 4C\,.\,(r^2 - a^2).$$

## 2.8 Conclusions

The ingredients for suitable plate theories have been identified in this chapter. These are equilibrium equations and deformation (kinematic) relations. The two simplest and most useful coordinate systems have been considered, namely rectangular cartesian and radially symmetric polar. In the next two chapters these ingredients will be blended to provide useful plate theories.

# 3 Elastic plates

## 3.0 Introduction

The earlier chapters have dealt with statics and kinematics of plate bending. That is, they have dealt with the conditions which the internal forces and moments must satisfy if they are to be in equilibrium with the loads; also, the methods have been developed for describing the deformation of the plate in its bent state. Before actual problems of plate bending can be formulated and solved, these two aspects have to be brought together, and in this chapter this is done for the case of the elastic plate. Roughly speaking, plates respond elastically in the working load condition, and plastically at the ultimate or collapse state. This chapter will be dealing with plates at working load, and the next will deal with plastic plates, namely plates in the final stages of loading before they collapse.

All the plates considered in this chapter will be assumed to be of constant thickness and isotropic, i.e. material properties not varying with direction in the plate.

## 3.1 Elastic theory of plate bending—moment/curvature relations

For simplicity, the coordinate system to be used will be assumed to be rectangular cartesian.

An elastic plate is characterized by a linear relation between load and deflection. However, the material response is not so much related to *deflection* as to *curvature change*. Recall the elastic beam, where the material response is a linear relation between bending moment and curvature change. The same is true for the plate, though now there is a two-dimensional aspect to the problem. The plate under load will develop internal force systems of bending moments (including twisting moments) and transverse shear forces to resist the loading and the appropriate equilibrium equations must also be satisfied.

In addition, the plate will deform, and the originally plane plate will deflect

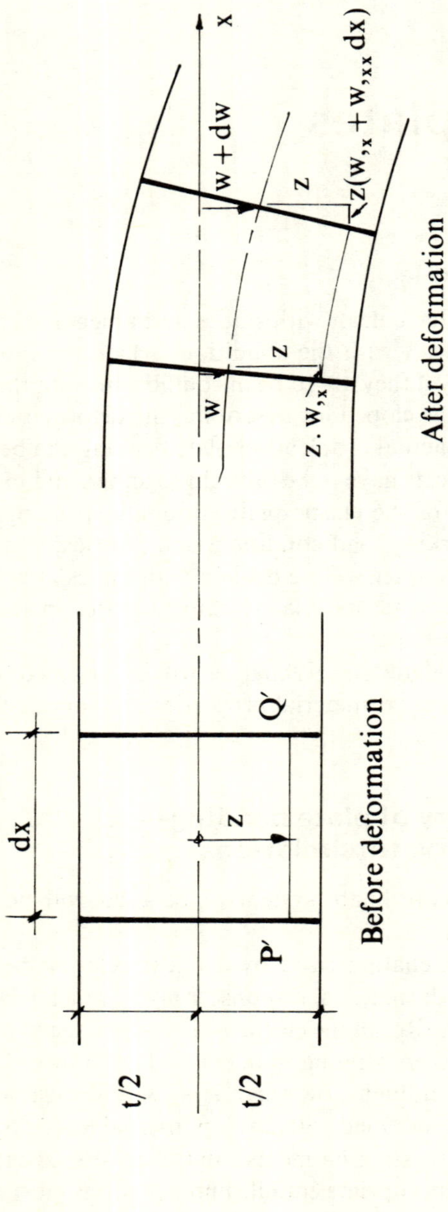


**Figure 3.0** Deformation geometry.

and develop curvatures and twists. For an elastic plate in bending, the bending moments relate linearly to the curvatures. The deformation of the plate has been characterized by the deformation of the middle surface. In the plate, the material on one side of the middle surface will be compressed, that on the other will be extended.

The plates with which we are concerned will be assumed to deform according to the *Kirchoff hypothesis.* This is the name associated with a mechanical model of behaviour, in which the plate can be thought of as composed of a series of laminae, as in plywood. One of these, the middle one, is the middle surface which can deform by bending but is not extended or compressed in its own plane. If the other plys are thought of as secured by nails through the thickness, then the Kirchoff hypothesis states that these nails are not extended or sheared by the deformation. Expressed in other words, the plane sections taken normal to the middle surface before deformation remain plane, and normal to the middle surface and unextended after deformation (Fig. 3.0).

Consider then the analysis of this model of behaviour. The requirement is to evaluate the strain in the x-direction at transverse distance z from the unstrained middle surface. If the transverse displacement is denoted by $w(x, y)$ then the slope in the x-direction will be $w_{,x}$ and

$$\begin{aligned}\varepsilon_{xx} &= \frac{z(w_{,x}+w_{,xx}\,dx)-z(w_{,x})}{dx}\\ &= zw_{,xx},\ \text{here compression.}\end{aligned} \tag{3.1.0}$$

In a similar manner

$$\varepsilon_{yy} = zw_{,yy} \quad \text{and} \quad \varepsilon_{xy} = zw_{,xy}.$$

But from the earlier discussion of middle surface curvatures it is seen that

$$\varepsilon_{xx} = z\kappa_{xx}, \varepsilon_{xy} = z\kappa_{xy}, \varepsilon_{yy} = z\kappa_{yy}. \tag{3.1.1}$$

Now the elastic response in this typical ply of the plate is

$$E\,.\,\varepsilon_{xx} = Ez\kappa_{xx} = \sigma_{xx} - \nu\sigma_{yy} - \nu\sigma_{zz} \tag{3.1.2}$$

and to a good approximation $\sigma_{zz} = 0$. (Here E is Young's modulus).

Recalling the definitions for stress resultants suggests that (3.1.2) should be multiplied by z and then an integration through the plate thickness performed.

Thus

$$\int_{-t/2}^{t/2} E\,.\,z^2\,.\,\kappa_{xx}\,dz = \int_{-t/2}^{t/2} \sigma_{xx}\,.\,z\,dz - \nu\int_{-t/2}^{t/2} \sigma_{yy}\,.\,z\,dz.$$

Then noting that $\kappa_{xx}$ is not a function of z,

$$\frac{Et^3}{12}\kappa_{xx} = M_{xx} - \nu M_{yy}.$$

In a similar manner

$$\frac{Et^3}{12}.\kappa_{yy} = M_{yy} - \nu M_{xx}.$$

Now solve for $M_{xx}$ and $M_{yy}$; then on multiplication by $\nu$ and adding, $M_{yy}$ can be eliminated and $M_{xx}$ found.

Hence

$$M_{xx} = D(\kappa_{xx} + \nu\kappa_{yy}), D \equiv \frac{Et^3}{12(1-\nu^2)},$$

and $$M_{yy} = D(\kappa_{yy} + \nu\kappa_{xx}).$$

In a similar manner, for each ply being sheared

$$2G.\varepsilon_{xy} = \sigma_{xy}$$

where G is the shear modulus.

By a similar process there is obtained

$$M_{xy} = \frac{2Gt^3}{12}\kappa_{xy} = D(1-\nu).\kappa_{xy}.$$

Hence collecting the elastic material relations together gives

$$\begin{bmatrix} M_{xx} \\ M_{yy} \\ M_{xy} \end{bmatrix} = D\begin{bmatrix} 1 & \nu & 0 \\ \nu & 1 & 0 \\ 0 & 0 & 1-\nu \end{bmatrix}\begin{bmatrix} \kappa_{xx} \\ \kappa_{yy} \\ \kappa_{xy} \end{bmatrix}. \quad (3.1.3)$$

*Notes*: The relations (3.1.3) summarize the elastic material response. They relate bending and twisting moments to *changes* in curvature and twist for the deformed middle surface.

*Exercise:* Find the bending and twisting moments associated with the deflected shapes defined in the two examples in 2.5. *Answer:*

$M_{xx} = -2AD[y(b-y)+\nu x(a-x)]$, $M_{xy} = AD(1-\nu)(a-2x)(b-2y)$.

## 3.2 Elastic theory of plate bending—the governing equation (cartesian coordinates)

The ingredients in the elastic theory of plate bending are the moments/unit length $M_{xx}$, $M_{yy}$, $M_{xy}$ (3 quantities), the transverse shear forces/unit length $Q_x$, $Q_y$ (2 quantities) and the middle surface changes of curvature and twist

$\kappa_{xx}$, $\kappa_{yy}$, $\kappa_{xy}$ (3 quantities). These eight quantities are dependent, and are generally unknown, and have to be found by solving the particular problem. The loading p(x, y) will be assumed specified, and all quantities are functions of x and y.

Eight equations are required to connect the eight dependent quantities, but there is a further very important dependent quantity, w(x, y), the transverse displacement, to be considered, so there are nine dependent quantities in all, and thus nine equations are required: these are three equilibrium equations, three moment-curvature relations and three relations expressing curvatures in terms of the displacement, w.

Now in (2.4.3) it has been shown that plate bending equilibrium can be expressed by the single equation

$$M_{xx,xx} + 2M_{xy,xy} + M_{yy,yy} = p. \tag{3.2.0}$$

If the relations (3.1.3) are substituted for the stress resultants then (3.2.0) becomes

$$D[(\kappa_{xx} + \nu\kappa_{yy})_{,xx} + 2(1-\nu)\kappa_{xy,xy} + (\kappa_{yy} + \nu\kappa_{xx})_{,yy}] = p. \tag{3.2.1}$$

Recalling that $\kappa_{xx} = w_{,xx}$, $\kappa_{xy} = w_{,xy}$, $\kappa_{yy} = w_{,yy}$, then (3.2.1) can be written entirely in terms of w as

$$D[w_{,4x} + \nu w_{,2x2y} + 2(1-\nu)w_{,2x2y} + w_{,4y} + \nu w_{,2x2y}] = p.$$

The $\nu$ terms cancel, whence

$$w_{,4x} + 2w_{,2x2y} + w_{,4y} = p/D. \tag{3.2.2}$$

This equation in the single dependent variable w is usually written in shorthand as

$$\nabla^4 w = p/D. \tag{3.2.3}$$

This is the celebrated *Biharmonic equation* and is *the governing equation* for elastic plate bending analyses.

*Notes*: (1) When comparing plate theory with beam theory it can be seen that there is a formal resemblance between

$$EIw^{IV} = p \quad \text{(see (1.6.6))},$$

and

$$D\nabla^4 w = p \quad \text{(see (3.2.3))}.$$

Note also the formal similarity between (3.2.0) and (1.6.2).

Indeed, this resemblance is quite useful to recall from time to time, and is an aid to anticipating plate behaviour by comparing with beam equivalents. The main differences stem from the twisting action which can be developed by plates but not by beams.

(2) The biharmonic is a *linear* equation and hence superposition of solutions is possible.

## 3.3 Circular plates—radial symmetry

The governing equation (3.2.3), the biharmonic equation, has been derived using the rectangular cartesian coordinate system. There will be occasions when other coordinate systems are needed, particularly the polar system. Here the plate equations for the simplest polar system, with radial symmetry, will be described.

In the previous chapter the equations of equilibrium were derived, so too were expressions for the curvatures (2.6, 2.7). The curvatures, when allied with the Kirchoff hypothesis of plane sections remaining plane give the link between forces on the one hand and strains and deformations on the other.

Thus the direct strain in the r-direction at a transverse distance z from the middle plane is given by

$$\varepsilon_{rr} = z\,.\,\kappa_{rr} = zw_{,rr}. \tag{3.3.0}$$

Again, in the transverse direction

$$\varepsilon_{\theta\theta} = z\,.\,\kappa_{\theta\theta} = z\frac{w_{,r}}{r}.$$

There is no shear strain because of the assumed radial symmetry. These strains then are those experienced by a layer of the plate parallel to the middle surface. As earlier in this chapter, using 3.3.0 in the elastic relations then multiplying by z and integrating through the thickness, the required stress resultants are recovered and suitable moment-deformation relations are derived.

Instead of this fundamental approach here it is observed that when this procedure is carried out in two dimensions the following type of relations result. If the directions of the coordinates are denoted by $(..)_1$ and $(..)_2$, and the coordinates are orthogonal, then

$$M_{11} = D(\kappa_{11} + \nu\kappa_{22}),$$

and (3.3.1)

$$M_{22} = D(\kappa_{22} + \nu\kappa_{11}),$$

where

$$D = \frac{Et^3}{12(1-\nu^2)},$$

E = Young's Modulus, $\nu$ = Poisson's Ratio and t is the plate thickness.

Here then

$$M_{rr} = D\left(w_{,rr} + \nu \frac{w_{,r}}{r}\right),$$
$$M_{\theta\theta} = D\left(\frac{1}{r} w_{,r} + \nu w_{,rr}\right). \tag{3.3.2}$$

Now take stock of what dependent variables are present, and how many equations relate. The dependent quantities are $M_{rr}$, $M_{\theta\theta}$, $Q_r$; $\kappa_{rr}$, $\kappa_{\theta\theta}$ and w; six in all. The available equations are two equilibrium (2.6.2), two stress-strain (3.3.2) and two curvature displacement (2.7.2). Hence there are six equations and six unknowns. The preferred unknown, as for rectangular coordinates, is w, the transverse displacement. The equations are

$$\begin{aligned} (rM_{rr})_{,r} - M_{\theta\theta} &= r\,.\,Q_r, \\ (rQ_r)_{,r} &= p\,.\,r, \\ M_{rr} &= D(\kappa_{rr} + \nu\kappa_{\theta\theta}), \\ M_{\theta\theta} &= D(\kappa_{\theta\theta} + \nu\kappa_{rr}), \\ \kappa_{rr} = w_{,rr}, \kappa_{\theta\theta} &= \frac{1}{r} w_{,r}. \end{aligned} \tag{3.3.3}$$

The equations are structured in such a way that the three forces and moments are described by *only two* equilibrium equations—that is, the circular plate is a statically indeterminate problem.

The simplest procedure for solution is to derive the analogous equation to (3.2.3); indeed this equation will be (3.2.3) expressed in polar coordinates. To achieve this, first eliminate $Q_r$ between the equilibrium equations, thus

$$(rM_{rr})_{,rr} - M_{\theta\theta,r} = p\,.\,r. \tag{3.3.4}$$

Then eliminate $M_{rr}$, $M_{\theta\theta}$ by use of the moment-curvature relations to obtain

$$(r\kappa_{rr} + \nu\kappa_{\theta\theta})_{,rr} - (\kappa_{\theta\theta} + \nu\,.\,\kappa_{rr})_{,r} = \frac{pr}{D}. \tag{3.3.5}$$

Finally, substitute the values of the curvatures in terms of the displacement derivatives, when

$$(rw_{,rr} + \nu w_{,r})_{,rr} - \left(\frac{1}{r} w_{,r} + \nu w_{,rr}\right)_{,r} = \frac{pr}{D}.$$

Simplifying by expanding terms and cancelling, when

$$(rw_{,rrr} + w_{,rr})_{,r} + \nu(w_{,rrr}) - \frac{1}{r}w_{,rr} + \frac{1}{r^2}w_{,r} - \nu w_{,rrr} = \frac{pr}{D},$$

or

$$w_{,4r} + \frac{2}{r}w_{,3r} - \frac{1}{r^2}w_{,2r} + \frac{1}{r^3}w_{,r} = \frac{p}{D}. \qquad (3.3.6)$$

This equation can be written in another form which is more easily used to construct a solution. Thus (3.3.6) can be grouped and then written as

$$\frac{1}{r}\left[r\left(\frac{1}{r}(rw_{,r})_{,r}\right)_{,r}\right]_{,r} = \frac{p}{D}. \qquad (3.3.7)$$

This second form then allows an integration in quadratures, that is, repeated direct integration.

*Note*: As remarked earlier, (3.3.6 and 7) are in reality (3.2.3) expressed in polar coordinates but with all variation with respect to $\theta$ suppressed.

## 3.4 Some simple solutions for circular plates

Following up the suggestion in the previous section, (3.3.7) will be integrated by repeated integration. Thus multiply through by r and then integrate with respect to r, when

$$r\left(\frac{1}{r}(rw_{,r})_{,r}\right)_{,r} = \frac{pr^2}{2D} + \bar{A}.$$

Now divide by r and carry out a further integration, when

$$\frac{1}{r}(rw_{,r})_{,r} = \frac{pr^2}{4D} + \bar{A}\ln r + \bar{B}.$$

Next multiply by r and integrate a third time, then

$$(rw_{,r}) = \frac{pr^4}{16D} + \bar{A}\left[\ln r.\frac{r^2}{2} - \int \frac{r^2}{r}.\frac{1}{r}dr\right] + \frac{\bar{B}r^2}{2} + C.$$

Finally

$$w = \frac{pr^4}{64D} + \frac{\bar{A}r^2}{8}[2.\ln r - 1] + \frac{\bar{B}r^2}{2} + \bar{C}.\ln r + \bar{D}.$$

Defining new arbitrary constants, which is a convenience rather than a necessity, the general solution of (3.3.6) and (3.3.7) can be seen to be

$$w = \frac{pr^4}{64D} + A.\ln r + Br^2.\ln r + Cr^2 + E. \qquad (3.4.0)$$

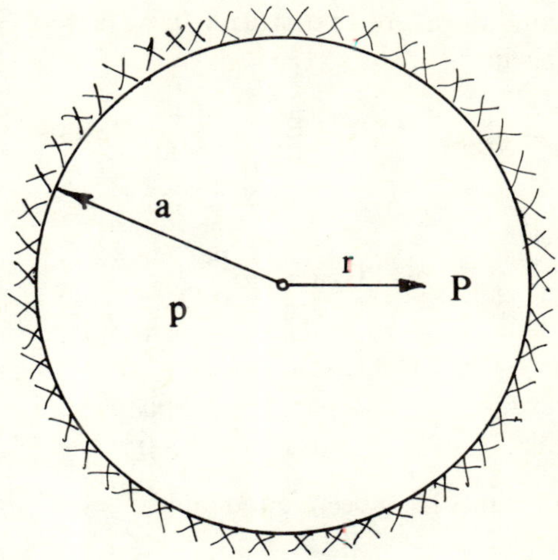


**Figure 3.1** Circular clamped plate—uniform loading.

This solution consists of a complementary function, namely the portion containing the four arbitrary constants A, B, C and E together with the *particular integral* for p, which has been evaluated here on the assumption that p is a constant. If p is some function of r then suitable particular integrals are readily calculated, exactly in the manner of the present one, if required. Also, as for beams, so here, loadings such as point loads cannot be directly represented in the solution, since they represent discontinuous happenings in the member, and (3.4.0), being the regular solution of an ordinary differential equation (3.3.7), is written explicitly for *full continuity*. Discontinuities can be handled but this is a specialized point which will not be considered here.

Consider an example of a uniform, clamped-edge circular plate, radius a, with a uniform distribution of applied loading, p (Fig. 3.1). Find an expression for the deflected shape and evaluate the bending moments. The starting point is the general expression for the transverse displacement, (3.4.0).

Hence

$$w = \frac{pr^4}{64D} + A\,.\ln r + Br^2\,.\ln r + Cr^2 + E.$$

As with beam problems, so here the search for conditions to fix the values for the (four) arbitrary constants must be sought from two conditions at each end of the range of integration. The range of integration here is $0 \leqslant r \leqslant a$.

At $r = 0$ the requirement is that w be finite and $w_{,r} = 0$. These conditions can be met only if $A = B = 0$, since $\ln r$ is singular at $r = 0$. These then are the two conditions at $r = 0$.

At $r = a$, the conditions are that $w(a) = 0$, $w_{,r}(a) = 0$. From the general expression for w, then

$$C.a^2 + E = -\frac{pa^4}{64D},$$

and

$$2C.a = -\frac{pa^3}{16D}.$$

Hence

$$C = -\frac{pa^2}{32D}, \quad E = \frac{pa^4}{64D}.$$

Thus all four constants have been found and

$$w(r) = \frac{pr^4}{64D} - \frac{pa^2r^2}{32D} + \frac{pa^4}{64D}$$

$$= \frac{p}{64D}(a^2 - r^2)^2. \tag{3.4.1}$$

This then is the shape taken up by the plate for a uniform loading.

To evaluate the bending moments, it is convenient to first find the curvatures from (3.4.1).

Now

$$w_{,rr}(=\kappa_1) = \frac{p}{16D}[-a^2 + 3r^2],$$

and

$$\frac{1}{r}w_{,r}(=\kappa_2) = \frac{p}{16D}[-a^2 + r^2].$$

Hence

$$M_{rr} = D\left(w_{,rr} + \nu\frac{1}{r}w_{,r}\right) = +\frac{pr^2}{16}(3+\nu) - \frac{pa^2}{16}(1+\nu),$$

and

$$M_{\theta\theta} = D\left(\frac{1}{r}w_{,r} + \nu w_{,rr}\right) = +\frac{pr^2}{16}(1+3\nu) - \frac{pa^2}{16}(1+\nu).$$

For small r both $M_{rr}$ and $M_{\theta\theta}$ are negative, and hence are *sagging* bending moments. Near the edge both moments are positive and hence are *hogging* bending moments. At $r = \sqrt{(1+\nu)/(3+\nu)}\,.\,a$, $M_{rr}$ changes sign, while at $r = \sqrt{(1+\nu)/(1+3\nu)}\,.\,a$, $M_{\theta\theta}$ changes sign.

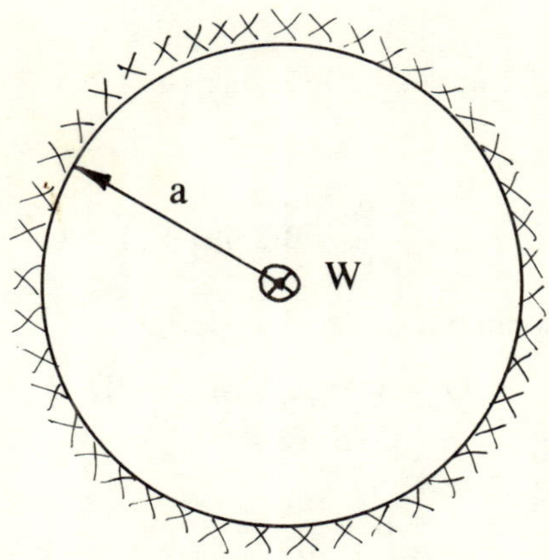


**Figure 3.2** Example, circular clamped plate—point loading.

*Notes*: (1) *Sagging* bending moment causes tension on the underside of the slab, *hogging* bending moment produces top tension.

(2) The form of the governing equation (3.3.6) can be integrated without knowledge of the form (3.3.7), but it is essentially a trial process. For example, suppose the solution is sought in the form $w = Ar^n$, then the terms in $r^2$ and a constant can be found, but not the other two involving $\ln r$.

(3) If the radius $r = 0$ is contained in the plate, that is if there is no hole at the centre, then for finiteness at $r = 0$, $A = B = 0$ must follow. No further conditions at $r = 0$ can then be applied—for example a zero slope requirement from symmetry. Such conditions are implied by $A = B = 0$, and the remaining conditions at the outer radius.

As a second example consider the clamped circular plate but loaded with a central point load W rather than a uniform distribution of loading (Fig. 3.2). Find the deflected shape, and distribution of bending moments.

The procedure is to regard the point load as being distributed over a small circular area at the centre of the plate. Suppose the radius of this area to be $\delta$. The distribution of transverse shear force $Q_r$ then can be found from that which equilibrates this load.

Thus,

$$2\pi\delta Q_r = W.$$

Actually, $\delta$ can be any radius r and it follows that $rQ_r = W/2\pi$. Since $p \equiv 0$ in this case,

$$w = A.\ln r + Br^2.\ln r + Cr^2 + E.$$

Then

$$w_{,r} = \frac{A}{r} + B(2r\,.\ln r + r) + 2Cr$$

and

$$w_{,rr} = -\frac{A}{r^2} + B(2\,.\ln r + 3) + 2C.$$

Now $M_{rr}$ and $M_{\theta\theta}$ can be evaluated, and so

$$r\,.\,Q_r = (rM_{rr})_{,r} - M_{\theta\theta} = 4D\,.\,B.$$

Hence

$$B = \frac{rQ_r}{4D} = \frac{W}{8\pi D}.$$

The remaining conditions are slope = 0 at the origin,

$$(A = 0),\ w(a) = 0 \quad \text{and} \quad w_{,r}(a) = 0.$$

Hence

$$E = \frac{Wa^2}{16\pi D}, \quad C = -\frac{W}{16\pi D}(2\ln a + 1).$$

Finally,

$$w(r) = \frac{W}{8\pi D}\left[r^2\,.\ln\frac{r}{a} - \frac{1}{2}(r)^2 + \frac{a^2}{2}\right].$$

Hence the central deflection is

$$w(0) = \frac{Wa^2}{16\pi D}.$$

*Exercise:* Consider the plate of 3.4, but now with the edge simply supported rather than clamped. The conditions at the (outer) edge will then be that the displacement and radial bending moment are zero. For simplicity assume Poisson's ratio ($\nu$) to be zero, and show that

$$w(r) = \frac{p}{64D}(a^2 - r^2)(5a^2 - r^2),$$

$$M_{rr}(r) = -\frac{3p}{16}(a^2 - r^2),$$

$$M_{\theta\theta}(r) = -\frac{p}{16}(3a^2 - r^2).$$

Can you devise a check on these results by use of the clamped edge results?

## 3.5 Simple solutions for problems in rectangular coordinates

The quest now is for solutions of the biharmonic equation, (3.2.3). The simplest useful problems relate to rectangular plates.

A basic method of solution and really the only feasible formal method for solution of engineering plate problems, is the method of *separation of variables*. There are other methods, considered later, but these are essentially numerically based, or approximate methods.

The method of separation of variables relates mostly to plates of rectangular shape, and begins by assuming that w(x, y), the dependent quantity being sought, can be written as a *product* of two separate functions of X(x) and of Y(y).

Thus the solution process begins by letting

$$w(x, y) = X(x) . Y(y). \tag{3.5.0}$$

Usually this is not a sufficiently strong assumption to allow the solution to proceed. A stronger assumption is therefore

$$w(x, y) = X(x) \operatorname{Sin} \frac{n\pi y}{b} \tag{3.5.1}$$

which is a relevant assumption to make for a rectangular plate, simply supported along y = 0, b.

Consider the simply supported rectangular plate, Fig. 3.3. Suppose the pressure loading to be sine varying in each of the x and y directions as indicated.

Then the solution of $\nabla^4 w = p(x, y)/D$ can proceed by *assuming* that w is given by

$$w = A . \operatorname{Sin} \frac{\pi x}{a} \operatorname{Sin} \frac{\pi y}{b}. \tag{3.5.2}$$

The motivation is that the solution sought should be of the separated variables type *and* should satisfy the edge conditions. This latter requirement is that

$$\begin{aligned} & 0 = w(x, 0) = w(0, y) = M_{yy}(x, 0) = M_{xx}(0, y), \\ \text{and} \quad & 0 = w(x, b) = w(a, y) = M_{yy}(x, b) = M_{xx}(a, y). \end{aligned} \tag{3.5.3}$$

These conditions look rather formidable. What they imply is that the simple support conditions must be satisfied. The conditions are the plate equivalent of the ordinary beam simple support conditions, namely that the displacement and the moment *about* the edge line should be zero, all round the boundary.

Thus w(x, 0) = 0 is the requirement that the edge displacement along the

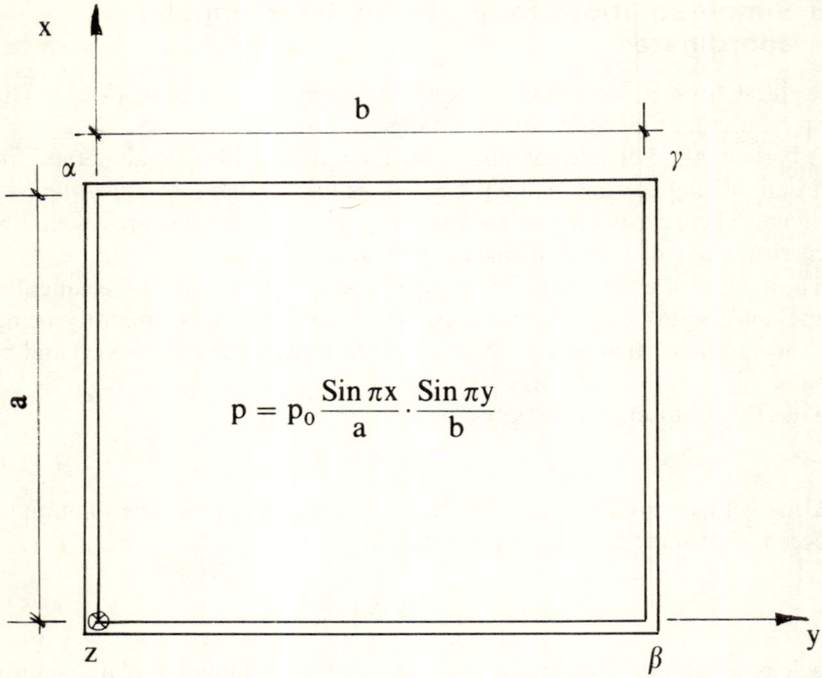


**Figure 3.3** Rectangular plate, simple support.

edge $0\alpha$ should be zero. The associated moment condition along this edge is that $M_{yy}(x, 0) = 0$, since the simple support condition requires that the plate rest on the edge, supported so as to ensure that $w(x, 0) = 0$ but not restrained rotationally, and hence $M_{yy}(x, 0) = 0$.

Now it can be seen that the choice of an x variation in (3.5.2) of $\sin(\pi x/a)$ is the only one available which will ensure the satisfaction of these edge conditions, for $\sin(\pi x/a)$ goes to zero at $x = 0, a$ and hence ensures that $w(0, y) = w(a, y) = 0$.

In addition, along the edge $0\beta$, $\kappa_{yy} = 0$ by virtue of $w(0, y) = 0$, and hence

$$\begin{aligned} M_{xx}(0, y) &= D[\kappa_{xx}(0, y) + \nu\kappa_{yy}(0, y)] \\ &= D \,.\, w_{,xx}(0, y). \end{aligned}$$

Now since w has been assumed to have an x variation of $\sin(\pi x/a)$, then this is sufficient to ensure that $w_{,xx}(0, y) = 0$ if $w(0, y) = 0$.

Hence it can be confirmed that *all* the edge conditions (3.5.3) are indeed satisfied by the assumed form of w variation (3.5.2). This is the essential importance of the separation of variables form for w, coupled with the individual variations being Sine or Cosine functions (depending upon where the origin of coordinates is positioned). (It should also be added that the importance of these features for the simply supported boundary also contributes to the near uselessness of the separation of variables in most other physically interesting edge conditions.)

To proceed with the solution, there remains the single task of ensuring that the assumed displacement form (3.5.2), which contains an unknown amplitude, A, is able to satisfy the governing biharmonic equation. Making the substitution of (3.5.2) into (3.2.2), and recalling the particular p(x, y) proposed here, gives

$$A\left[\left(\frac{\pi}{a}\right)^4+2\left(\frac{\pi}{a}\right)^2\left(\frac{\pi}{b}\right)^2+\left(\frac{\pi}{b}\right)^4\right]\operatorname{Sin}\frac{\pi x}{a}\operatorname{Sin}\frac{\pi y}{b}=p_0\operatorname{Sin}\frac{\pi x}{a}\operatorname{Sin}\frac{\pi y}{b}.$$

Making a comparison of coefficients it is seen that

$$A=\frac{p_0}{\left[\left(\frac{\pi}{a}\right)^2+\left(\frac{\pi}{b}\right)^2\right]^2}$$

Then

$$w=\left(\frac{ab}{\pi}\right)^4\cdot\frac{p_0}{(a^2+b^2)^2} \tag{3.5.4}$$

is an exact solution to this particular problem.

Some space has been devoted to this rather special example because it displays in a simple way all the essential features of a separation of variables solution. This particular example is of the simplest type, and the solution consists of a single, simple expression. Most examples encountered in practice are likely to be summations of terms derived from superposition of a series of components of the loading expression, which has been expressed as a double Fourier series in order to allow the solution to proceed along the lines adopted above.

The essential features of the process are, first, consideration of the boundary conditions and what form the separation of variables solutions must take in order to satisfy as many of the boundary conditions as possible, and then a formal solution of the governing equation by substitution and comparison of coefficients.

## 3.6 Further separation of variable features—rectangular plates

In the previous section the simplest solution of all for a rectangular plate was discussed, and the technique of separation of variables was adopted. In this section this same technique will be used, but in a more general context.

Separation of variables usually implies that at least one coordinate variation can be taken to be of Sine or Cosine type. Indeed, the usefulness of the technique depends upon such a variation in association with a pair of parallel simply supported edges normal to the associated coordinate direction. This is because such a combination can be made to satisfy the boundary conditions on the pair of simply supported edges. Once this is achieved, the problem can be reduced to an ordinary differential equation in the other variable, and this is a much simpler problem than the original one. The process described here is the so called *Levy method* of solution.

Consider the following example—a square plate, simply supported on one parallel pair of edges, fixed on the remaining pair, with a uniform loading (Fig. 3.4).

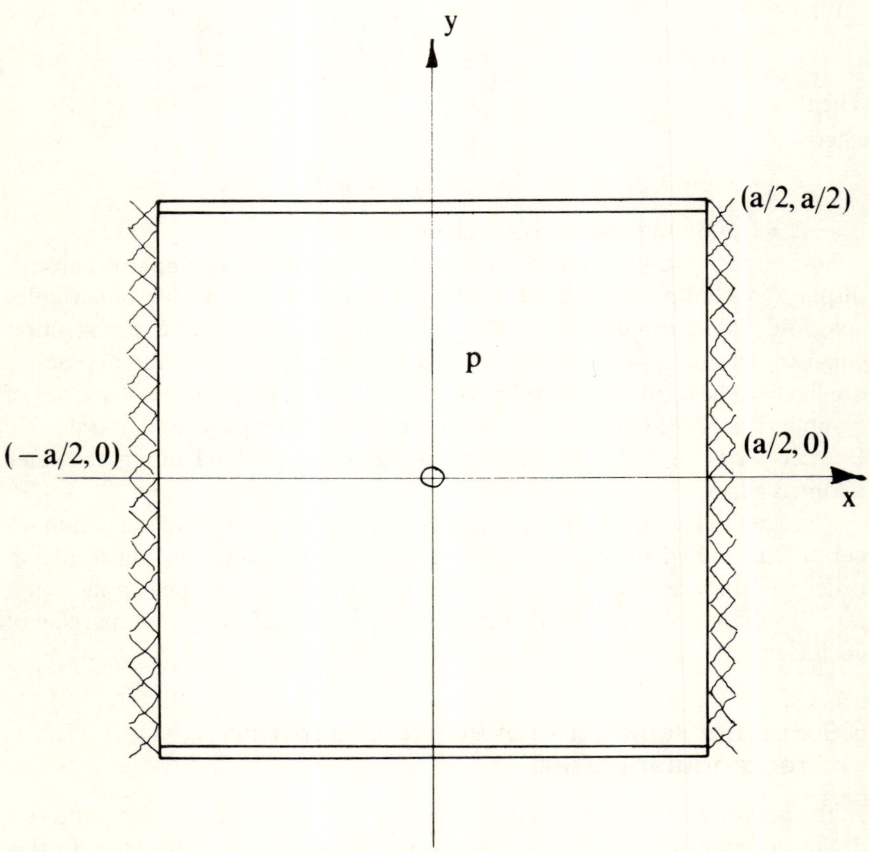


**Figure 3.4** Rectangular plate, mixed edge conditions.

According to the (Levy) procedure described above, the simply supported edge conditions are satisfied by a variation in the y direction of Cos $n\pi y/a$.

Hence

$$w = \sum \phi_n(x) . \text{Cos}\frac{n\pi y}{a} \tag{3.6.0}$$

is a suitable form for the displacement when n is odd.

It is easy to check that the edge conditions

$$w\left(x, \pm\frac{a}{2}\right) = M_{yy}\left(x, \pm\frac{a}{2}\right) = 0$$

are satisfied by *each* term in the assumed form for w, (3.6.0). Thus the simply supported edge conditions are implied by (3.6.0). It remains to construct the value for $\phi_n(x)$ such that the fixed edge conditions should also be satisfied.

First express the pressure loading as a Fourier Cosine series, thus

$$p_0 = \sum b_n . \text{Cos}\frac{n\pi y}{a},$$

where

$$b_n = \frac{4p_0}{\pi} . \frac{(-)^{\frac{n+3}{2}}}{n} \tag{3.5.1}$$

with n odd.

Now use (3.6.0) and (3.6.1) in the governing equation (3.2.2) when the following series of ordinary differential equations for $\phi_n$ is obtained:

$$\phi_n^{IV}(x) + (-2)\left(\frac{n\pi}{a}\right)^2 . \phi_n^{II}(x) + \left(\frac{n\pi}{a}\right)^4 . \phi_n(x) = \frac{4p_0}{\pi D} . \frac{(-)^{\frac{n+3}{2}}}{n} \tag{3.6.2}$$

where n is odd, and the individual equations for $\phi_n(x)$ have been obtained by separately equating to zero the coefficients of the respective terms in Cos $(n\pi y/a)$.

The dominant term in the solutions of the problem is likely to be that for $n = 1$. Hence proceed with the analysis for this term. The equation (3.6.2) is of constant coefficient type, and solutions should be sought of the form $w = Ae^{\alpha x}$. Using this in (3.6.2), and setting the right-hand side to zero for the present, the following equation for $\alpha$ is obtained:

$$\alpha^4 - 2\left(\frac{\pi}{a}\right)^2 . \alpha^2 + \left(\frac{\pi}{a}\right)^4 = 0. \tag{3.6.3}$$

Then

$$\left[\alpha^2-\left(\frac{\pi}{a}\right)^2\right]^2 = 0, \quad \alpha = \pm\left(\frac{\pi}{a}\right), \text{ repeated.}$$

Hence the basic components in the complementary function solution are Sinh $\pi x/a$ and Cosh $\pi x/a$. The repeated root indicates that x times these same components are also solutions. Finally, a suitable particular integral for the pressure can be easily constructed by noting that a $\phi_p(x) = \text{const.} = K$ say, gives $\phi_p(x) = 4p_0a^4/\pi^5$.

Hence the general solution of (3.6.2) for n = 1 is

$$\phi_1(x) = (A+Bx)\,\text{Cosh}\frac{\pi x}{a} + (C+Ex)\,\text{Sinh}\frac{\pi x}{a} + \frac{4a^4}{\pi^5}\cdot\frac{p_0}{D}. \qquad (3.6.4)$$

In the present example, $\phi_1(x)$ must be *even* in x, from the symmetry about the x = 0 line.

Hence B = C = 0. There remain the boundary conditions on x = a/2 to be satisfied, then those on x = −a/2 will be satisfied by symmetry.

Hence, at x = a/2

$$w\left(\frac{a}{2}, y\right) = \phi_1\left(\frac{a}{2}\right).\,\text{Cos}\frac{\pi y}{a}$$

$$= \left(A\,.\,\text{Cosh}\frac{\pi}{2} + E\,.\,\frac{a}{2}\text{Sinh}\frac{\pi}{2} + \frac{4a^4}{\pi^5}\cdot\frac{p_0}{D}\right)\text{Cos}\frac{\pi y}{a}$$

$$= 0$$

and

$$w_{,x}\left(\frac{a}{2}, y\right) = \frac{d\phi_1}{dx}.\,\text{Cos}\frac{\pi y}{a}$$

$$= \left(A\,.\left(\frac{\pi}{a}\right).\,\text{Sinh}\frac{\pi}{2} + E\left[\text{Sinh}\frac{\pi}{2} + \frac{\pi}{2}\text{Cosh}\frac{\pi}{2}\right]\right).\,\text{Cos}\frac{\pi y}{a} = 0.$$

Solving

$$E = -1.16\frac{A}{a} \quad \text{and} \quad A = \frac{-3.41}{\pi^5}.\frac{p_0a^4}{D}. \qquad (3.6.5)$$

Then

$$(w_{max})_{n=1} = w(0,0) = A + E\,.\,0 + \frac{4a^4}{\pi^5}\cdot\frac{p_0}{D}$$

$$= 0.0211\frac{p_0a^4}{Et^3}, \text{ (using } \nu = 0.3). \qquad (3.6.6)$$

The complete solution, using all the odd values of n as implied by the Fourier series (3.6.1), can be shown to give $w_{max} = 0.0209\,p_0a^4/Et^3$.

The bending moments can also now be calculated, and for example,

$$M_{xx}\left(\pm\frac{a}{2},0\right) = D\,(w_{,xx} + \nu w_{,yy}]_{a/2,0}$$

$$= D\left[\left(\frac{\pi}{a}\right)^2 A\,\mathrm{Cosh}\frac{\pi}{2} + 2E\frac{\pi}{a}.\mathrm{Cosh}\frac{\pi}{2} + E\frac{a}{2}\left(\frac{\pi}{a}\right)^2 \mathrm{Sinh}\frac{\pi}{2}\right]$$

$$= 0.075\,.\,p_0a^2,\ \text{hogging.} \qquad (3.6.7)$$

This value compares with the complete solution value of

$$M_{xx}\left(\frac{a}{2},0\right) = 0.070\,.\,p_0a^2.$$

*Exercise:* Show that the above values for $n = 1$ case give $M_{xx}(0,0) = -0.035p_0a^2$, $M_{yy}(0,0) = -0.028p_0a^2$. The exact values for a uniform load, including all Fourier components for n, are

$$M_{xx}(0,0) = -0.032pa^2, \quad M_{yy}(0,0) = -0.024pa^2.$$

A negative moment indicates a sagging value, tension on underside.

## 3.7 Solution by finite differences

Thus far only analytical solutions for the plate equation (3.2.2) have been considered. In many situations of practical interest, analytical solutions are not attainable, usually because the boundary conditions do not conform to a type which can be handled by separation of variables methods. Hence numerical solutions must be sought.

A very suitable technique is the *finite difference method*. Here the aim is to evaluate the transverse displacement at a regular grid of points across the plate, rather than construct functional expressions for the displacement.

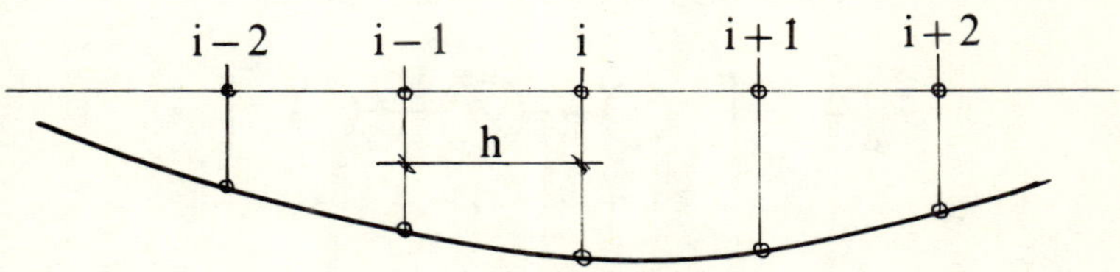


**Figure 3.5** Difference grid—one dimension.

When set up, the problem becomes an algebraic one of solving a set of simultaneous, algebraic equations for the values of w at the chosen grid of points. Thus the differential equation aspect of the problem is removed by the approximation procedure. The basis of the method is as follows.

Consider first a one-dimensional situation and let there be a regular set of nodal points spaced h apart (Fig. 3.5). For definiteness, Fig. 3.5 can be considered to be a section through a deflected plate. The aim is to relate the values of w at the nodal points $i-2$, $i-1$, $i$, $i+1$, $i+2$ etc. to the derivatives of w at the central point i.

Now $w_{i+1}(x)$ can be expanded as a Taylor series as follows

$$w_{i+1}(x) = w_i(x) + h\frac{dw_i(x)}{dx} + \frac{h^2}{2!}\frac{d^2w_i(x)}{dx^2} + \ldots$$

Likewise

$$w_{i-1}(x) = w_i(x) - h\frac{dw_i(x)}{dx} + \frac{h^2}{2!}\frac{d^2w_i(x)}{dx^2} - \ldots$$

If h can be considered to be small, say in relation to the total span of the plate, then it should be legitimate to truncate these Taylor series and solve for $dw_i/dx\,(x)$ in terms of nodal values of $w_i$. If terms in $h^2$ and higher powers of h are neglected, then solving for $dw_i/dx$ gives

$$\frac{dw_i(x)}{dx} = \frac{1}{2h}[w_{i+1}(x) - w_{i-1}(x)]. \tag{3.7.0}$$

This is a typical finite difference expression, where on the left-hand side is a sought derivative, and on the right, the appropriate array of nodal values of w(x).

Proceeding thus, the second derivative can be found, by neglecting terms in $h^3$ and beyond, to be

$$\frac{d^2w_i(x)}{dx^2} = \frac{1}{h^2}[w_{i+1}(x) + w_{i-1}(x) - 2w_i(x)]. \tag{3.7.1}$$

This array will be represented by the following "module",

$$h^2 . \frac{d^2w_i(x)}{dx^2} =$$

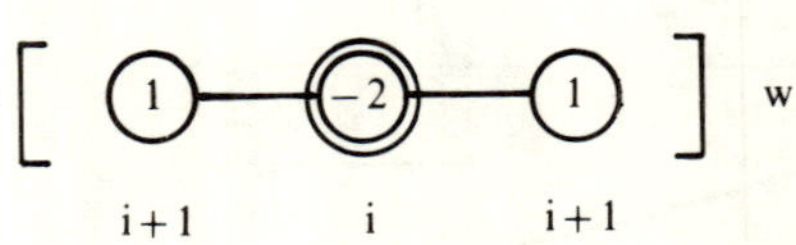


where the subscripts will often be omitted, but the meaning will be clear from the context.

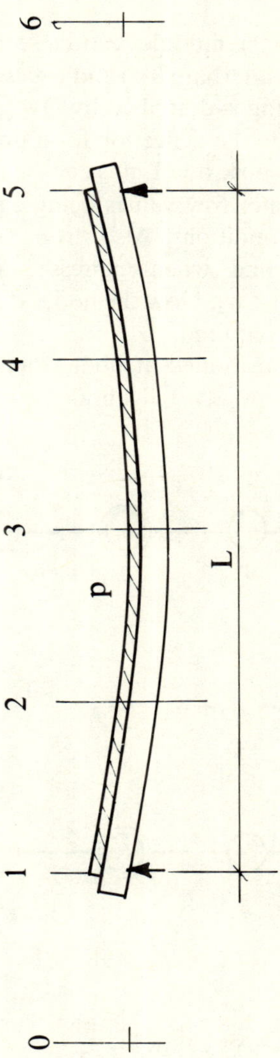


**Figure 3.6** Example—difference grid on beam.

Again

$$h^4 . \frac{d^4 w_i(x)}{dx^4} =$$ 

(3.7.2)

Indeed, the coefficients in the modules can be seen to be given by rows in a Pascal triangle array, with alternate (+) and (−) signs.

For example, beam bending is described by $Dw_{,xxxx}(x) = p$. Use the finite difference method to estimate the deflection for a uniform uniformly loaded, simply supported beam. Suppose $h = L/4$ is chosen (Fig. 3.6). Then there are 5 nodal points 1–5, and hence 5 w values being sought. But $w_1 = w_5 = 0$, from symmetry and end conditions. Also, from the remaining boundary condition, $M_1 = M_5 = 0$, and when expressed in w terms this gives $w_4 + w_6 - 2w_5 = 0$, or $w_6 = -w_4$. Now the nodes 0, 6 are imaginary, though their notional values will be required.

If the module (3.7.2) is positioned at node 2 and then node 3, with the symmetry requirement $w_4 = w_2$, two algebraic equations for $w_2$ and $w_3$ are obtained.

Thus

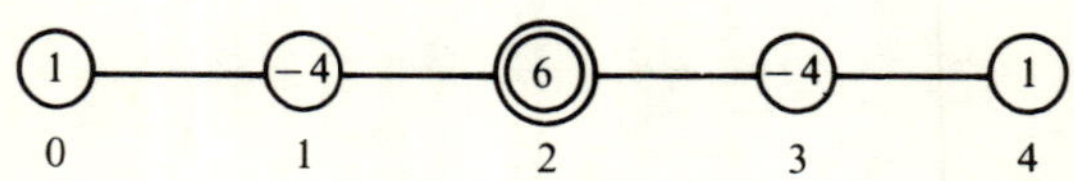


$= ph^4/D$

gives
$$6w_2 - w_2 + w_2 - 4w_3 = \frac{ph^4}{D},$$

and

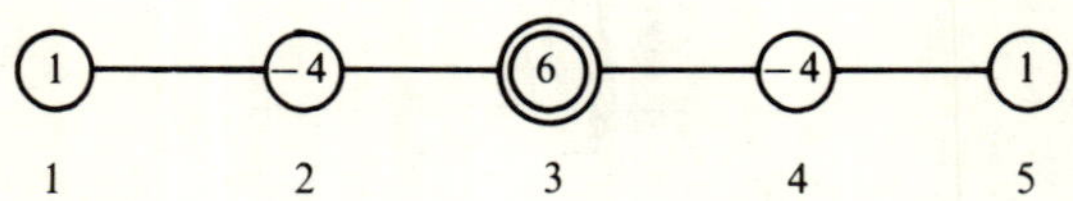


$= ph^4/D$

gives
$$-8w_2 + 6w_3 = \frac{ph^4}{D},$$

or

$$\begin{bmatrix} 6 & -4 \\ -8 & 6 \end{bmatrix} . \begin{bmatrix} w_2 \\ w_3 \end{bmatrix} = \frac{ph^4}{D} . \begin{bmatrix} 1 \\ 1 \end{bmatrix}, \begin{bmatrix} w_2 \\ w_3 \end{bmatrix} = \frac{ph^4}{4D} \begin{bmatrix} 6 & 4 \\ 8 & 6 \end{bmatrix} . \begin{bmatrix} 1 \\ 1 \end{bmatrix}.$$

Hence

$$w_3 = \frac{14}{4} \cdot \frac{p\left(\frac{L}{4}\right)^4}{D} = \frac{7pL^4}{512D} = 0.00136\frac{pL^4}{D}.$$

The exact solution is of course

$$w = \frac{5}{384}\frac{pL^4}{D} = 0.00130pL^4/D.$$

This result is good although only a crude mesh of points was used.

*Exercise:* Repeat the previous example, but with a central point loading, W, instead of a pressure load. Show that $w_{max} = 0.0234\,WL^3/D$ for $h = L/4$, compared with the exact value of $0.0208\,WL^3/D$.

Plate problems require that the governing equation (3.2.2) be approximated by a suitable finite difference module. The terms such as $w_{,4x}$, $w_{,4y}$ have been

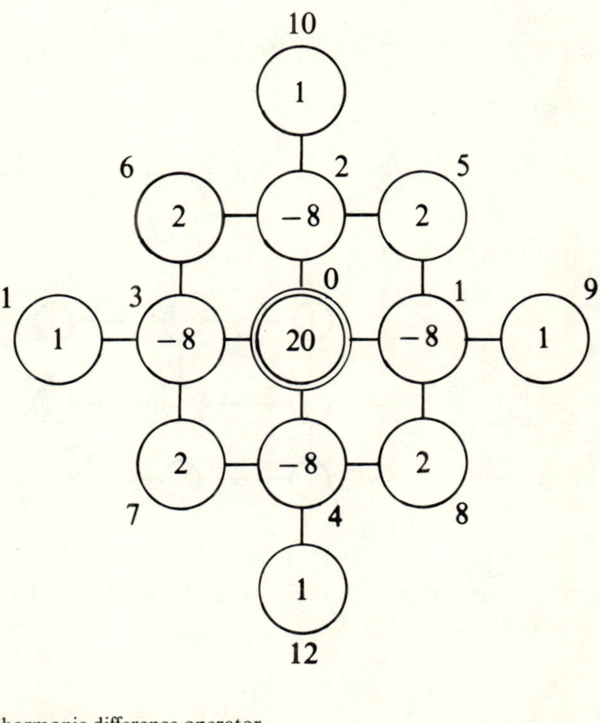


**Figure 3.7** Biharmonic difference operator.

derived (3.7.2). There remains to construct the module for $w_{,2x2y}$. This can be carried through by formal manipulation of the Taylor series expansion in two independent variables. However, it is preferred here to argue physically with the use of the module for the second derivative (3.7.1).

Consider the array of nodal points (Fig. 3.7). The object is to evaluate $w_{0,2x2y}$.

Now

$$h^2 w_{1,2y} = w_5 + w_8 - 2w_1,$$

$$h^2 w_{0,2y} = w_2 + w_4 - 2w_0,$$

$$h^2 w_{3,2y} = w_6 + w_7 - 2w_3,$$

and

$$h^2(\ )_{0,2x} = (\ )_1 + (\ )_3 - 2(\ )_0.$$

Hence

$$h^2(h^2 w_{0,2y})_{,2x} = (w_5 + w_8 - 2w_1) + (w_6 + w_7 - 2w_3) - 2(w_2 + w_4 - 2w_0)$$

Simplify, when

$$h^4 . w_{0,2x2y} = (w_5 + w_6 + w_7 + w_8) - 2(w_1 + w_2 + w_3 + w_4) + 4w_0. \quad (3.7.3)$$

This information is represented as the module

$$h^4 \frac{d^4 w_0}{dx^2 dy^2} =$$

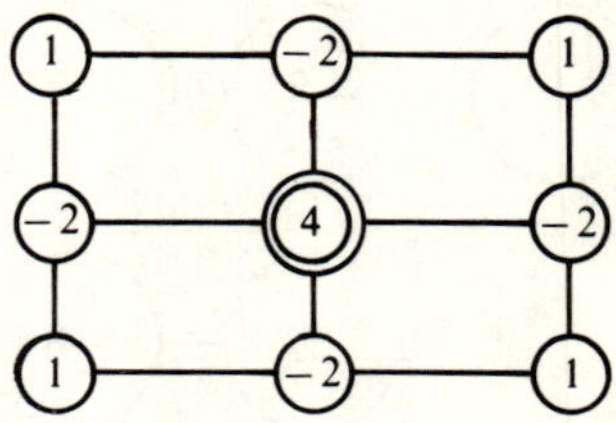


Now collect the three terms in (3.2.2) when the final complete module for the bi-harmonic becomes

$$w_{,4x} + 2w_{,2x2y} + w_{,4y} = \frac{p}{D}$$

or

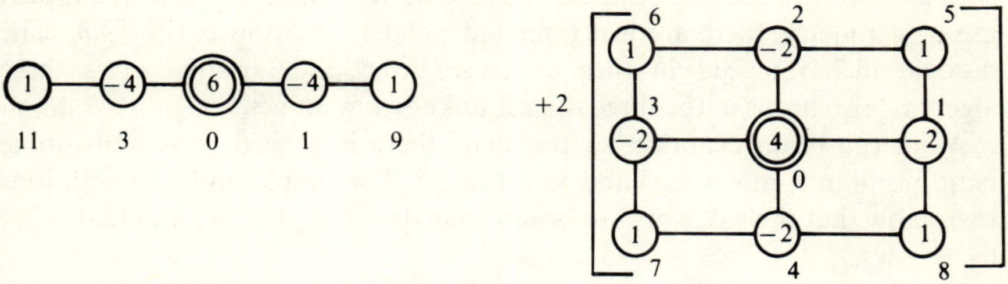


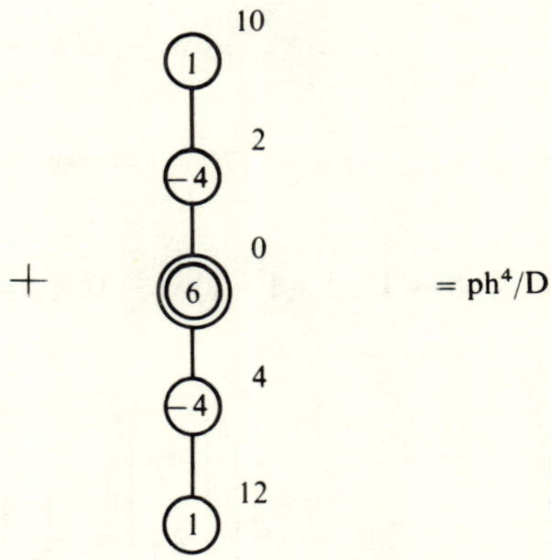


Namely

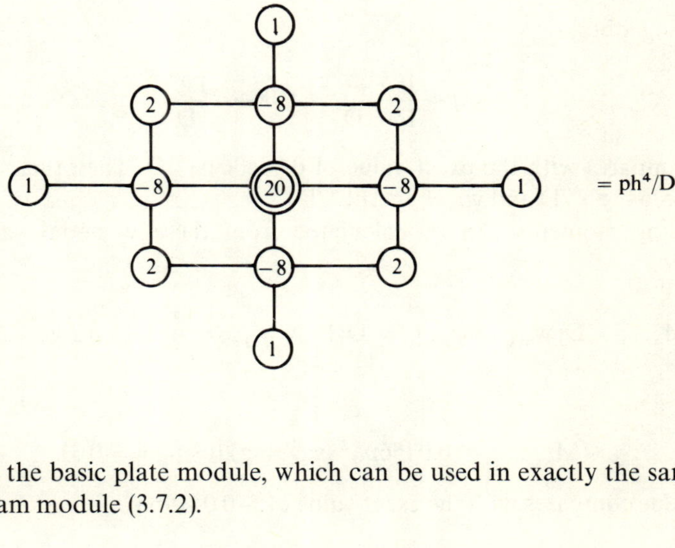


(3.7.4)

This is the basic plate module, which can be used in exactly the same way as the beam module (3.7.2).

For example, consider a square, simply supported plate (side a) loaded with a uniformly distributed load, p. To begin with, take $h = a/4$. Then, with use of symmetry, there are just three independent unknowns (Fig. 3.8). The module (3.7.4) placed in turn at these three nodal points gives three algebraic equations in the three sought unknown w values.

As in the beam example, so too here, there is a need to identify some fictitious points, shown as 3 and 4 on Fig. 3.8. The simple support conditions then show that the value of w to be associated with $w_4$ is $-w_1$ and that of $w_3$ to be $-w_2$.

Hence, with the module centred at 0, 2, 1 the following three equations are obtained:

$$20w_0 + 8w_1 - 32w_2 = \frac{ph^4}{D}$$

$$-8w_0 - 16w_1 + (20+1-1)w_2 = \frac{ph^4}{D}$$

$$2w_0 + (20+1+1-1-1)w_1 - 16w_2 = \frac{ph^4}{D},$$

or

$$\begin{bmatrix} 20 & +8 & -32 \\ 2 & 20 & -16 \\ -8 & -16 & 20 \end{bmatrix} . \begin{bmatrix} w_0 \\ w_1 \\ w_2 \end{bmatrix} = \frac{ph^4}{D} \begin{bmatrix} 1 \\ 1 \\ 1 \end{bmatrix}.$$

Solving, obtain

$$w_0 = \frac{165}{160}\frac{ph^4}{D} = 0.00402\frac{pa^4}{D}.$$

This compares with the exact value of $0.00406\,pa^4/D$. The other values are $w_1 = 35/64 \, . \, ph^4/D$ and $w_2 = 3/4 \, . \, ph^4/D$.

Bending moments can be calculated from these w nodal values. For example,

$$(M_{xx})_0 = D[w_{,xx} + \nu w_{,yy}]_0 = D(1+\nu)w_{,xx} = \frac{D}{h^2}(1+\nu)[2w_2 - 2w_0].$$

and

$$(M_{xx})_0 = -0.0456pa^2 \text{ (sagging) (using } \nu = 0.3).$$

This value compares with the exact value of $-0.0479pa^2$.

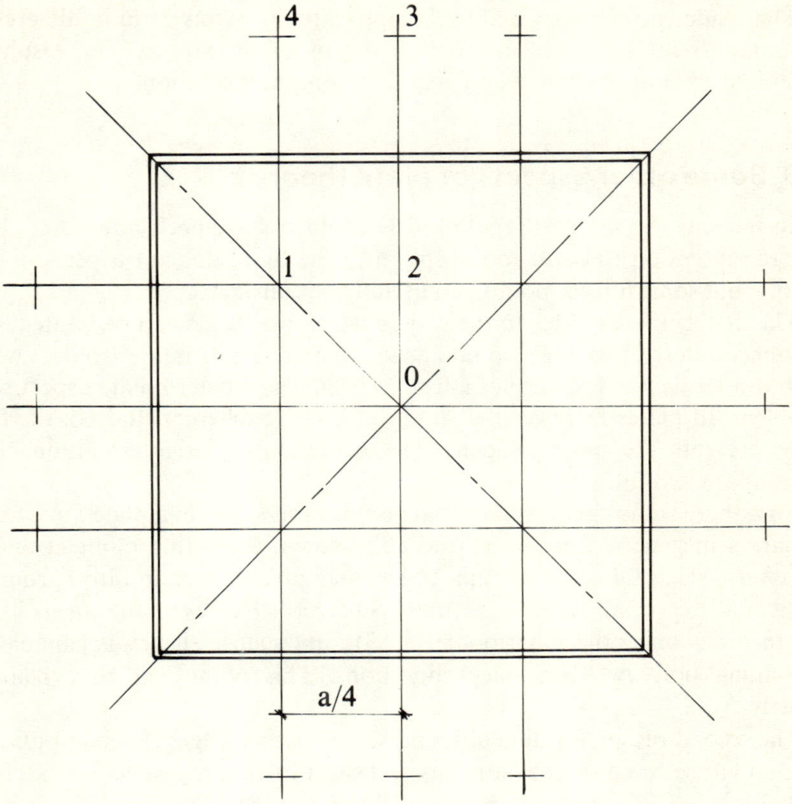


**Figure 3.8** Difference grid—simply supported plate

It must be admitted that the agreement, particularly on bending moment, from use with such a crude grid of points, is better here than can usually be expected.

*Exercises:* (1) An even cruder solution of this problem is possible, using $h = a/2$. Show that the estimate for $w_0$ in this case is $0.00390 . pa^4/D$, and that the maximum (sagging) moment is given by $-0.0393pa^2$ when $\nu = 0.3$.

(2) Use finite differences to solve, approximately, the example dealt with in Section 3.6—square plate simply supported on two and fixed on two other edges. Crudest estimate,

$$h = \frac{a}{2}, w_{max} = \frac{1}{20}\frac{ph^4}{D} = \frac{1}{320}\frac{pa^4}{D} = 0.0031\frac{pa^4}{D}.$$

For $h = a/4$, show $w_{max} = 0.0024\ pa^4/D$. The exact value is $0.0019\ pa^4/D$.

The reader now has available the apparatus in terms of finite difference modules to set up problems with many more unknowns. The resulting algebraic equations will however require a computer solution.

## 3.8 Some other aspects of plate theory

Thus far only the simplest types of plate problem have been considered. It is not the intention to delve too deeply into the more difficult aspects of the theory, but some further points can usefully be considered.

The first of these relates to the way in which free edges can be treated. On first encounter, a free edge would appear to be a simple matter to deal with. In beam terms, the presence of a free-end signifies a determinate aspect to a problem. In plates however the situation is quite different. Indeed, the free edge presents the most difficult type of elementary edge condition. The reasons are two-fold.

First there is the circumstance that at a transverse cut in a plate *three* force variables in general appear, a bending moment, a twisting moment and a transverse shear force. If this cut is to be a *free* edge then on intuitive grounds we would expect all three quantities to be zero. However, the *theory* is of *fourth* order only (the biharmonic (3.2.3)) and such a theory is capable of sustaining only *two* boundary conditions. This point will be explained shortly.

The second reason for difficulties presented by free edges is a manipulative one, and arises because the moments and shear forces are respectively second and third derivatives of the transverse displacement. The effect of this is to require boundary conditions to be imposed which are expressed as second and third derivatives of the displacement w. Thinking back to the processes required of a finite difference solution, it will be seen that such a requirement normally means more fictitious points are required, and this increases the problem complexity and hence the total difficulty.

To return to the question of three quantities at an edge and only two permissible conditions. On an examination of the edge $Q_x$ force and the twisting moment $M_{xy}$ on a face x = const. say which is either a supported or a free edge then it is concluded that these two quantities should be grouped to give a composite quantity

$$V_x = Q_x + \frac{\partial M_{xy}}{\partial y}. \tag{3.8.0}$$

This expression results from the observation made in (2.1). There it was noted that $M_{xy}$ can be visualized as a distribution of $Q_x$ type forces forming couples of strength $M_{xy}$. As $M_{xy}$ varies so there is a near cancellation of the $M_{xy}$ forces at any point seen from left and right. Only $\partial M_{xy}/\partial y$ remains, and hence the above expression (3.8.0) results.

Then at a *supported* edge $V_x$ supplies the vertical edge reaction while at a *free* edge $V_x = 0$ is one boundary condition to be coupled with the second condition $M_{xx} = 0$. Hence there are just *two* conditions.

It is not proposed to explore this question much further except to note that this combining of $Q_x$ and $M_{xy}$ leads to the appearance of concentrated corner forces, typically of value $2M_{xy}$ at a corner in a rectangular plate. The following advice is also offered: in a first look at a plate problem with a free edge, especially if some approximate solution such as a finite difference solution is being sought, then rather than approach the solution in a fully rigorous way, expect that in the neighbourhood of the free edge the plate will tend toward a *cylindrical* shape. By suitable interpretation of this tendency it is often possible to avoid introducing many, or any, fictitious points with a consequent simplification of the problem formulation.

It has been suggested already that most practical elastic plate problems, which are not standard problems, are likely to be solved numerically, since there is so little scope to solve these problems analytically. Hence a method of numerical solution must be decided upon, and today there is a strong likelihood that this choice will come down in favour of a Finite Element approach.

The essential aim of the *Finite Element method* (F.E.M.) is to reduce the continuum plate problem to one which can adequately be described by a *finite* number of parameters. This process is one of approximation and as with all approximation procedures, experience and judgement are needed to fully exploit the potential of the method.

The Finite Difference method described in the previous section is an approximation procedure which is essentially mathematically based. The difference calculus is a branch of classical mathematical analysis, and the only real choice open to the user of the method is the mesh size which is to be adopted.

The finite element method is a physically based approximation procedure, and the user of the method is faced with a large number of decisions in the formulation of the whole finite element calculation. Herein lies the scope for experience and judgement to play a part in achieving an acceptable solution.

In a work such as this, only the briefest of introduction can be given to this method. In essential features the F.E.M. solution uses a stiffness matrix formulation based on an *assumed* set of element characteristics. There are now many specialized accounts of the method, for example Chapter 5 of J. S. Przemieniecki's *Theory of Matrix Structural Analysis* (McGraw-Hill, 1968). In real situations it is becoming less likely that an individual user will

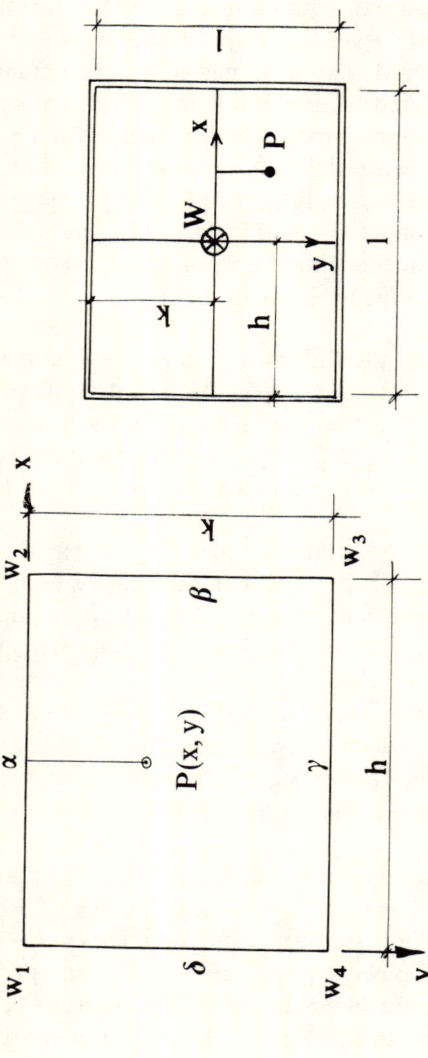


**Figure 3.9** Finite elements—element and example.

himself write the necessary computer program to carry through an F. E. M. analysis—it is more likely that an available bureau program will be used. Then the onus is on the recipient of the results to *check* the feasibility of what has been obtained, from the point of view of how well the F.E.M. model has been able to simulate the problem in hand. It is in this part of the problem exploration that some of the foregoing methods and results can be employed.

As a sample of the F.E.M. consider a rectangular bending element (Fig. 3.9*a*). Eight parameters, the four corner transverse displacements and the four side curvatures, will be taken to describe the element shape. Alone however these eight quantities cannot define w(x, y) internal to the element. Hence *assume* additionally that the curvatures vary *linearly* away from the edges in each direction. Then it follows that $w_p(x, y)$ at P is given by

$$w_p = w_1\left(1-\frac{x}{h}-\frac{y}{k}\right)+w_2\frac{x}{h}+w_4\frac{y}{k}+(w_1-w_2+w_3-w_4)\frac{xy}{hk}$$

$$+\frac{\alpha}{2}x(x-h)+\frac{\delta}{2}y(y-k)-\frac{\alpha}{2}\frac{yx}{k}(x-h)+\frac{\beta}{2}\frac{x}{h}y(y-k)$$

$$+\frac{\gamma}{2}\frac{y}{k}x(x-h)-\frac{\delta}{2}\frac{x}{h}y(y-k). \tag{3.8.1}$$

This is a very crude and not very useful shape, but it can serve to illustrate some points of the method. The shape automatically allows element displacement but not slope continuity. For example,

$$\nabla^4 w_p = 0, \tag{3.8.2}$$

showing that the element is incapable of supporting distributed loading. However, suppose the problem of a point loaded rectangular plate (Fig. 3.9*b*), is to be studied. Four elements with obvious symmetries can span the plate, though in any serious study many more would be used.

From the support conditions it is the case that for a typical element

$$w_2 = w_3 = w_4 = 0 = \beta = \gamma. \tag{3.8.3}$$

Then from remarks made earlier (2.1 and 3.8) it is to be expected that point corner forces are generated of value $2M_{xy}$. These summed at the central node will support the point load, W. Thus

$$(M_{xy})_1 = D(w_{,xy})_1 = \frac{Dw_1}{hk} \tag{3.8.4}$$

Hence

$$W = 4x2x\frac{Dw_1}{hk} \quad \text{or} \quad w_1 = \frac{WLl}{32D}. \tag{3.8.5}$$

This is a very poor estimate of w, around three times the accepted value, but serves to show how quickly certain information can be obtained by such approximation procedures.

The remaining parameters $\alpha$, $\delta$, the element edge curvatures, can be found from the slope continuity condition at the centre to be

$$\alpha = \frac{-2w_1}{h^2}, \quad \delta = \frac{-2w_1}{k^2}. \tag{3.8.6}$$

Now internal force information can be found. For example, for $\nu = 0$,

$$(M_{xx})_1 = Dw_{,xx} = D\alpha = \frac{-W}{4}\frac{l}{L}. \tag{3.8.7}$$

In effect this problem has a single displacement unknown, $w_1$. When many elements are used matrix methods are called for. Thus suppose the material response to be given by

$$\boldsymbol{\sigma} = \mathbf{De}, \tag{3.8.8}$$

where $\boldsymbol{\sigma}$, $\mathbf{e}$ are the internal stress and strain vectors, and $\mathbf{D}$ is the symmetric material response matrix—this stage corresponds to (3.1.3). Generally $\boldsymbol{\sigma}$ and $\mathbf{e}$ will vary from point to point. The vector of nodal displacements $\mathbf{d}$ and the strains $\mathbf{e}$ will be related through some chosen element properties, such as (3.8.1), and can be written symbolically as

$$\mathbf{e} = \mathbf{B}.\mathbf{d}. \tag{3.8.9}$$

Hence if the external forces $\mathbf{p}$ relate to the internal actions $\boldsymbol{\sigma}$ through a contragredient equilibrium relation of the type

$$\mathbf{p} = \int \mathbf{B}^T.\boldsymbol{\sigma}\,dV \tag{3.8.10}$$

where $(\ )^T$ is the matrix transpose and V is the volume of the element, then

$$\mathbf{p} = \int \mathbf{B}^T.\boldsymbol{\sigma}\,dV = \int \mathbf{B}^T.\mathbf{D}.\mathbf{e}\,dV = \int \mathbf{B}^T.\mathbf{D}.\mathbf{Bd}.dV$$

$$= \int \mathbf{B}^T.\mathbf{D}.\mathbf{B}\,dV.\mathbf{d} = \mathbf{K}.\mathbf{d}. \tag{3.8.11}$$

Here $\mathbf{K} = \int \mathbf{B}^T.\mathbf{D}.\mathbf{B}\,dV$ is the element stiffness matrix. By suitable assembly of these element stiffness matrices, a complete F.E.M. problem can be formulated and solved for the nodal displacements, $\mathbf{d}$. Thereafter the internal force information is obtained by use of (3.8.8, 9).

Solutions must be carefully scrutinized to ensure that they adequately solve the problem in hand. Looked at from a different point of view, every problem solved provides the exact solution to *some* problem, but usually not the one aimed for.

## 3.9 Stability of plates

Thus far the formulation of the elastic plate problem has been centred on the bi-harmonic equation (3.2.3). In this section it is proposed to supplement this equation with the necessary terms which will enable the *stability of equilibrium* of the elastic plate to be studied.

The essential differences between the plates studied in this section and those earlier in the chapter are, firstly, that equilibrium here is written down with explicit regard for the equilibrium being required in the *deformed* state of the plate and, secondly, that the typical loading which the plate here must withstand is forces of $N_{xx}$, membrane type, which act in the original plane of the plate.

The plate is still primarily being *bent*, but the bending is being brought about by in-plane forces acting on the transverse displacements which develop when the plate becomes unstable.

As has been remarked, the loading of interest when studying plate stability

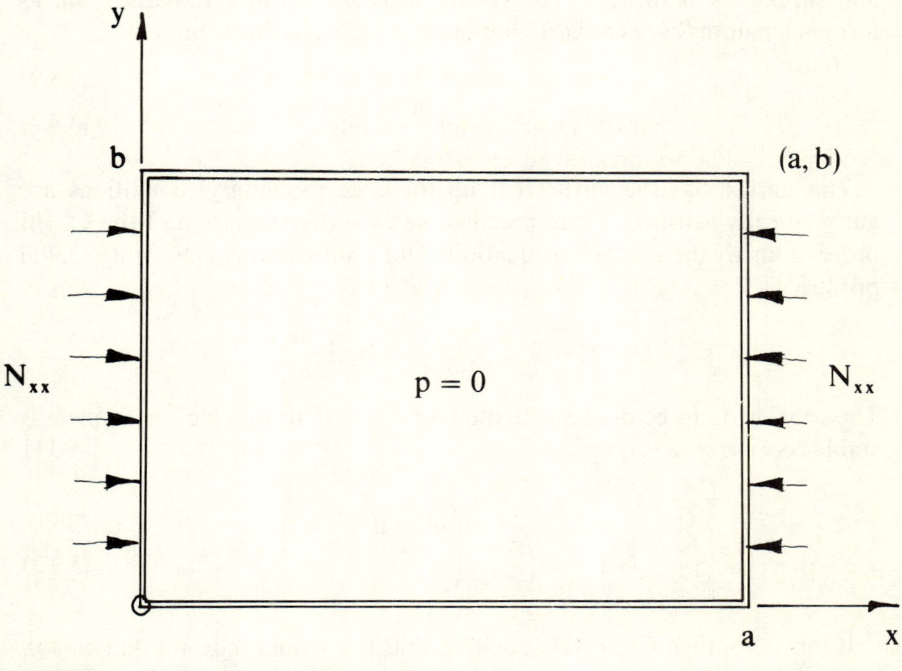


**Figure 3.10** Plate stability—uniaxial compression.

is from in-plane rather than out-of-plane forces. These forces of type $N_{xx}$ (Fig. 3.10) then can be seen to exert moments of the type $N_{xx}\,dw$ between the ends of a plate element. Hence the moment equilibrium equation about an axis parallel to the y axis becomes

$$M_{xx,x}+M_{xy,y}+N_{xx}w,_x = Q_x. \tag{3.9.0}$$

The $N_{xx}$ force is transmitted unchanged across the plate, because of the absence of other in-plane forces. Hence $N_{xx,x} = 0$.

Now when the single moment equilibrium equation is formed by eliminating the $Q_i$, then (3.9.0) contributes a term of the form $N_{xx}\,.\,w,_{xx}$. Finally, when the $M_{\alpha\beta}$ values are replaced by the curvature expressions, but expressed in terms of the transverse displacement, then the governing equation for the elastic plate being compressed in one direction is found to be

$$D\nabla^4 w+N_{xx}w,_{xx} = 0. \tag{3.9.1}$$

The plus sign in this equation is physically important and indicates that the equation describes an instability phenomenon. If a minus had been present, then no instabilities occur and physically the plate would be under stretching rather than compression forces.

Consider now a solution for (3.9.1), when the physical arrangement of plate and support is as in Fig. 3.10. Assume a separation of variables in *Navier* form of a pair of Sine type variations, one in each axis direction.

Hence

$$w(x, y) = C_{mn}\,.\,\mathrm{Sin}\frac{m\pi x}{a}\,\mathrm{Sin}\frac{n\pi y}{b}. \tag{3.9.2}$$

This choice has the virtue that all the edge (boundary) conditions are automatically satisfied. There remains only the determination of the $C_{mn}$ in order to satisfy the governing equation (3.9.1). Substituting (3.9.2) into (3.9.1) produces

$$C_{mn}\left[D\pi^4\left\{\left(\frac{m}{a}\right)^2+\left(\frac{n}{b}\right)^2\right\}^2-N_{xx}\left(\frac{m\pi}{a}\right)^2\right]=0.$$

The conclusion to be drawn is that either $C_{mn} \equiv 0$, in which case the plate is stable because $w \equiv 0$, or

$$N_{xx} = D\left(\frac{\pi}{a}\right)^2.\left(\frac{m^2+n^2\left(\frac{a}{b}\right)^2}{m}\right)^2. \tag{3.9.3}$$

If this is so, then $C_{mn} \neq 0$ is possible, and this would indicate that $w \neq 0$, hence the plate has become unstable.

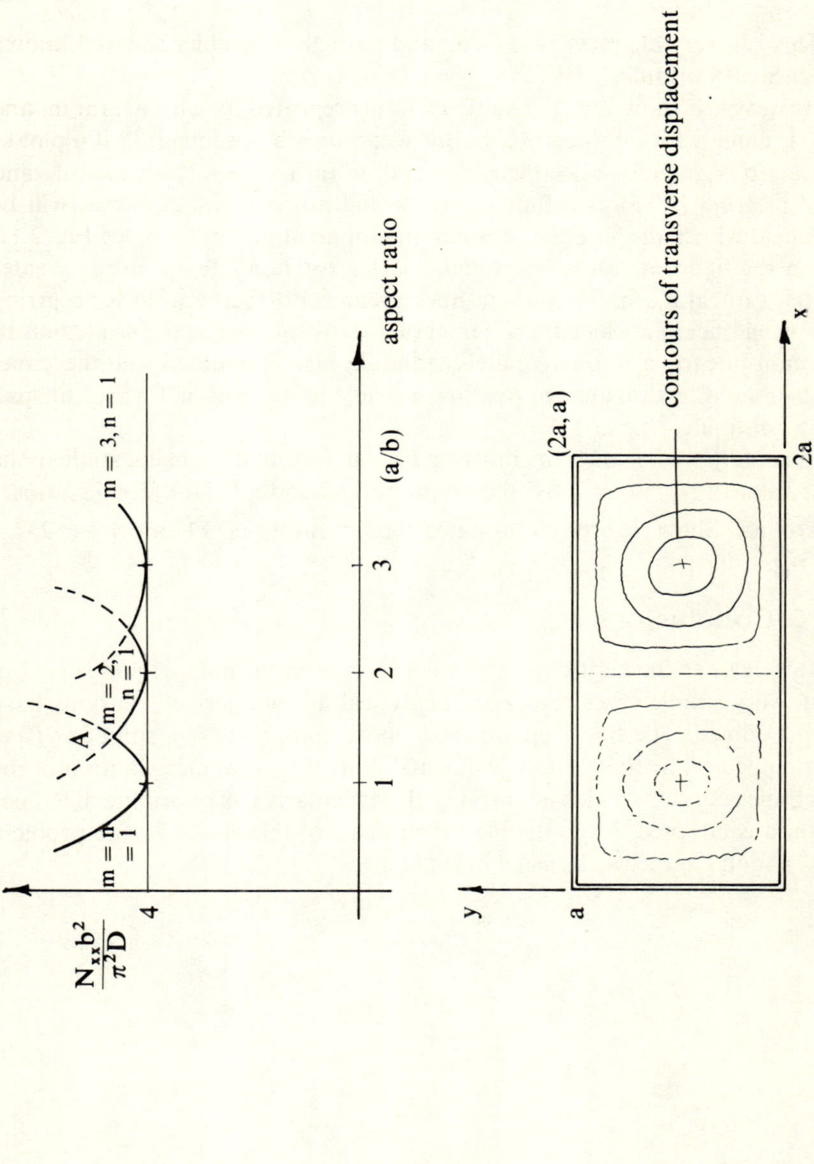


**Figure 3.11** Plate stability—typical results.

The parameters m, n must now be chosen to minimize $N_{xx}$. Suppose $a/b = 1$, then $N_{xx}$ is minimum for $m = n = 1$. For $a/b = 2$, the minimum $N_{xx}$ occurs for $m = 2$, $n = 1$. But in either case the minimum $N_{xx}$ is given by

$$(N_{xx})_{min} = \frac{4\pi^2 D}{b^2}, \quad b \leqslant a. \tag{3.9.4}$$

This is a very characteristic form, and strongly resembles the well-known Euler Strut Formula.

It is easy to show that the value (3.9.4) is repeated for all integral m, and $n = 1$, namely for all integral a/b. But what for a/b not integral? If a plot of N vs. a/b is made by selecting $n = 1$ and, in turn, integral values at m, and then plotting (3.9.3) as a function of N and a/b, a series of curves will be obtained which all share the common minimum at integral a/b, see Fig. 3.11. From the figure it can be seen that $(N_{xx})_{min}$ for non-integral a/b is greater than for integral a/b, though not much greater. Physically what is occurring is a panel buckling which for a/b integral means square panels of side $b(<a)$. For non-integral a/b square panels cannot be accommodated, and the waves in the long direction change to allow a whole number of half waves to span the whole plate.

For safety's sake, and remembering that an instability is being studied, the base value of $N_{xx} = 4\pi^2 D/b^2$ should probably be adopted for all side ratios.

*Exercise:* Show that the coordinates of point A on Fig. 3.11 are 4.5, $\sqrt{2}$.

## 3.10 Conclusions

In this chapter the elastic plate problem has been formulated and solved in some very simple cases, and both analytical and numerical solutions have been outlined. The first steps towards more comprehensive solutions have been aimed at. Such solutions will usually involve computer solution of the algebraic equations which are produced, particularly if (say) a finite difference method is adopted. More detailed treatments of this class of plate problem can be found in the works listed in chapter 6.

# 4 Plastic plates

## 4.0 Introduction

As remarked in Chapter 3, a plate near to collapse is likely to be in a plastic condition, whereas a plate at working load is likely to be in an elastic state. Compared with elastic plates, plastic response in plates is more varied, since it depends upon the nature of the plate material as to quite what response is to be expected.

In the first portion of this chapter solid metal plates, typically steel plates, will be studied. Later sections will deal with the response of reinforced concrete plates, but always in the near collapse, maximum load carrying capacity state.

Most plates at most stages of the loading cycle up to collapse have an important feature in common. This is the manner in which the strains are distributed through the plate thickness—the so-called *Kirchoff hypothesis*. In Chapter 3 this hypothesis is described in some detail. Here it is noted that this same *plane sections* behaviour is found to be a reasonable working hypothesis for plastic plates also. Thus the direct strains parallel to the plate surfaces are proportional to the distance from the middle surface at the section concerned.

## A. SOLID METAL PLATES

## 4.1 Yield criteria

In order to describe plate behaviour in the plastic condition, the material response must be defined. This is the role of the *yield criterion*. Consider a simple beam being bent, first of all elastically and then finally plastically (see Fig. 4.0). Initially there is a proportional, recoverable response along the portion OA of the moment/curvature plot. Then the response ceases to be elastic (A→B) and finally a plastic straining at constant moment (B→C) occurs. For a number of practical cases this full response can be approximated by the rigid-plastic response curve ODC. Such an idealization suggests

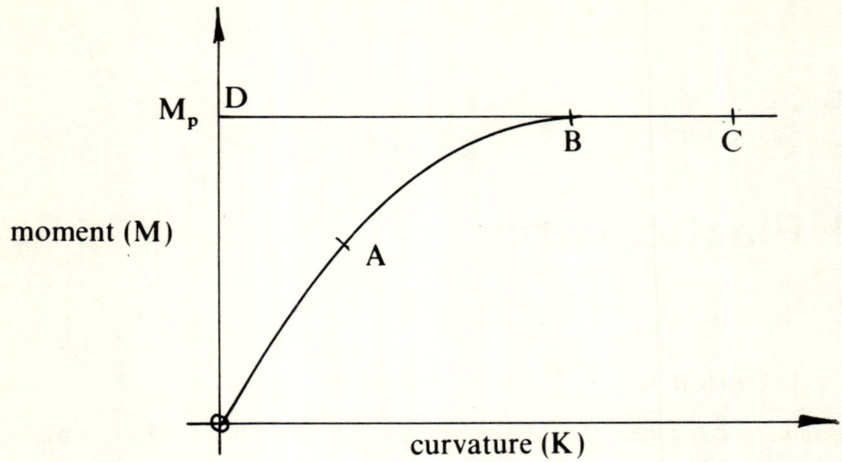


**Figure 4.0** Moment-curvature relations—beam element.

that the elastic curvatures are *small* when compared with the final curvatures and hence can be neglected compared with the plastic curvatures.

This material response data can be presented in another manner. For the beam described above the material response can be thought of as

$$0 > |M| > M_p, \tag{4.1.0}$$

that is, M is constrained to be numerically less than $M_p$, since the beam is unable to sustain a moment greater than $M_p$. The expression (4.1.0) is termed the yield criterion for the beam.

In the case of the plate, in general at every site in the plate there will exist a field of bending moments—in particular there will be two principal bending

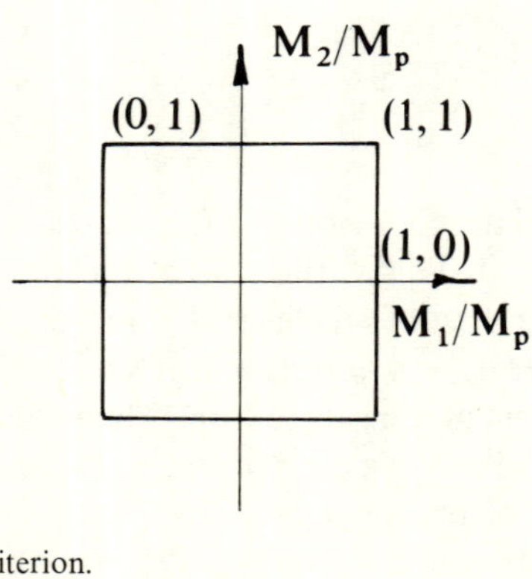


**Figure 4.1** Square yield criterion.

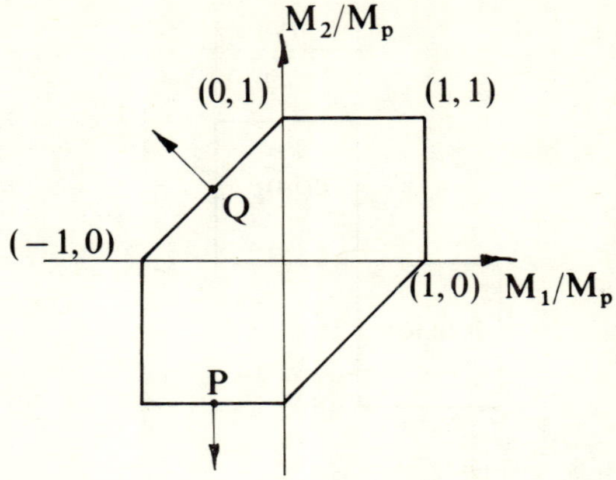


**Figure 4.2** Tresca yield criterion.

moments at each point. The yield criterion is most easily expressed in terms of principal moments.

Consider the yield criterion (Fig. 4.1). This is a principal moment plot, the so-called "square yield criterion", which is interpreted as meaning that either principal moment can rise to the one-dimensional full plastic value in either sense before yield is reached, and beyond which further moment cannot be sustained. There is some experimental evidence to suggest this is a reasonable criterion. In addition it is the simplest possible response for yielding in two dimensions.

The yield locus (Fig. 4.2) is the *Tresca Locus*, named after an early worker in this field. This locus is built up from the experimental observation that certain ductile materials begin to flow when the maximum shear stress reaches a critical value. As can be seen, the Square and Tresca criteria are essentially identical for moments of the same sign, but different when the moments are of opposing sign.

In either case the interpretation is that any combination of principal moments which plot to a point in the interior of the figures 4.1, 4.2 will be sustained without yield. If the combination plots to a point on the boundary of the figures, then yield is possible, while moment combinations which lie outside of the figures are unattainable. The two yield criteria so introduced are relevant for two-dimensional moment fields such as those encountered in plates being bent by transverse loads. Thought of at the level of an element of material, the square yield locus is an expression of the principal stress failure theory whereby each principal stress in the plane of the plate can rise to yield without influence from the other. By considering the plate to consist of a series of laminae, all of which on one side of the neutral (central) lamina

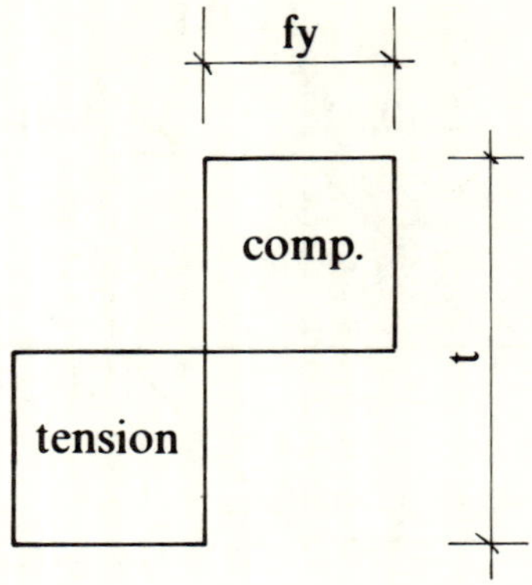


**Figure 4.3** Full plastic moments.

are stressed in one sense, say compression, while all those on the other are in tension, and integrating through the plate thickness after the manner of 2.1, so the moment resultants are obtained. If the yield stress is fy and the plate thickness t, then the moment resultant/unit plan length is given by $M = fy \, . \, t^2/4$. Hence $t^2/4$ is the per unit length, plastic section modulus.

In practical cases the square yield criterion will be used most extensively in the study of non-homogeneous reinforced concrete plates. (See Section B).

The Tresca criterion (Fig. 4.2) is more appropriate to solid homogeneous metal plates. If there is a need to consider the moment resultant in terms of the material yield stress, then the expression derived above ($M = fy \, . \, t^2/4$) applies here too. For yielding in the quadrants of Fig. 4.2 where the principal moments are of like sign, the larger principal stress coupled with the zero (or near zero) stress through the plate thickness produces yield by sliding along planes which are inclined at $\pi/4$ to the plate surface and intersect the plate surface along lines which are *parallel* to the *larger* principal moment vector.

By way of contrast, in those quadrants at Fig. 4.2 where the moments are of opposite sign, the (supposed) laminae of plate material are being *compressed* in one principal stress direction but *extended* in the other (Fig. 4.3). Now the maximum principal stress *difference* is not between the in-plane yield stress and the zero normal stress through the plate thickness, but is the maximum stress difference of 2fy between yield in tension and compression on orthogonal axes. The yielding now occurs on planes which are *normal* to the plate surface, namely planes directed through the plate thickness and equally inclined to the principal moment axes. This change in the planes of yielding has consequences for the type of deformation possible, and will be returned to below (4.3) where the question of curvature change and the Normality Rule are discussed.

## 4.2 The bound theorems

These theorems will be described and justified as physical consequences of the plastic material properties, and equilibrium. The importance of these theorems derives from the difficulty experienced in *fully* solving many problems of potential interest. Instead a restricted formulation of the problem, chosen because it can be solved, will form the basis of the solution, and then the bound theorems will be invoked to indicate whether the real solution implies a greater or lesser collapse load, for example.

In any complete problem there are essentially three component parts of the complete solution. On the one hand *equilibrium* must be achieved—in this case equilibrium at the limit of load-carrying capacity of the plate. Secondly, *yield* will have occurred at certain points or in certain regions of the plate; just where yielding has occurred must be investigated. Additionally the yield criterion must not anywhere be violated. Thirdly, the yield pattern must be such as to constitute a *mechanism* in the plate. The concept of a mechanism is geometrical and is a direct extension into two dimensions of the concept of a mechanism in a beam or framework. The basic requirement of a mechanism is that it should consist of patterns of lines, usually straight lines, seen in plan on the plate, *about* which the adjacent rigid regions of plate can rotate relative to one another. These lines of rotation are termed *yield lines* and must satisfy certain criteria if the whole pattern of lines is to form a mechanism.

For example, consider a square plate supported all round the perimeter (Fig. 4.4). Now if yield lines were to form across the two diagonals, then a *small* rotation of each quarter plate about the supported edge will cause points in a direction parallel to the $M_2$ axis. The normality rule then states produce a relative rotation between quarter plates to develop along the diagonals. The originally plane square plate will deflect into a square base (shallow) pyramid. The rotations must be considered small, and the deflections small compared with the plate span and thickness. With these

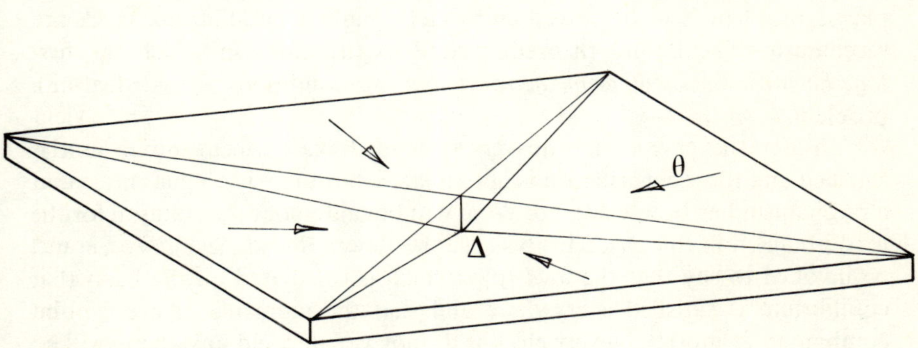


**Figure 4.4** Plate mechanism, one degree of freedom.

restrictions, the plate segments will not tend to separate across the yield lines. This is an example of a simple mechanism and will be called a mechanism with *one* degree of freedom.

The one degree of freedom can be taken to be $\Delta$ (Fig. 4.4), the central (maximum) displacement, and indicates that *one* displacement, say $\Delta$, can be chosen arbitrarily, whereupon the deflections of all other points in the plate middle surface follow for deformation of the plane middle surface into the pyramid shape.

A typical *two* degree of freedom mechanism might be a pyramid within a pyramid (Fig. 4.5). Now *two* typical displacements, say $\Delta$ and $\delta$, are required to determine the final displacements at all points.

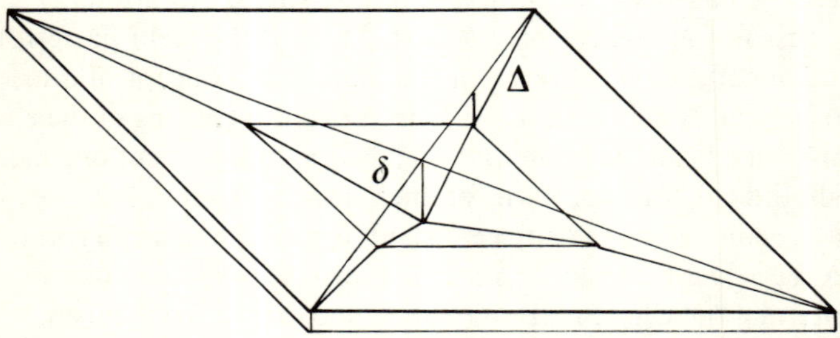


**Figure 4.5** Plate mechanism, two degrees of freedom.

In principle, mechanisms with very many degrees of freedom might occur in practice. However, in modelling this behaviour, minimum numbers of degrees of freedom will be assumed in order to keep the analysis as simple as possible, and this number will usually be restricted to one.

It has been said that three aspects must be simultaneously satisfied if the plastic problem is to be solved completely: namely equilibrium, yield and mechanism. The bound theorems relate to situations in which the first *together with one or other of the remaining* two conditions are satisfied in a problem.

Consider the case of a transversely loaded plate when equilibrium is satisfied and the yield criterion is not violated, but there is no guarantee that a mechanism has been achieved. What can be said about the solution for the load obtained in this case? In this case the *Lower Bound (Safe) theorem* will be invoked to say that the *load* ($p_L$), which is calculated on the basis that equilibrium is satisfied everywhere and that the moments at every point combine to at most achieve yield but do not violate yield anywhere, will be *less than* or at most *equal to* the true collapse load of the plate. The reason is that there is no guarantee that sufficient yield has occurred to form a mechanism, and hence the collapse load has not yet been reached.

On the other hand, suppose equilibrium is achieved and a mechanism is *imposed*, but no direct use of the yield criterion is made, then the *Upper Bound* (*Unsafe*) *theorem* says that the load ($p_u$) calculated this way is *greater* than or is perhaps *equal to* the true collapse load.

Hence we have that the true collapse load ($p_c$) is bounded above and below by $p_L$ and $p_u$, i.e.

$$p_L \leqslant p_c \leqslant p_u. \tag{4.2.0}$$

It is not proposed to offer formal proofs of these results. Instead, as opportunity arises, the correctness of (4.2.0) will be reinforced by examples.

In cases of true design, where the loads are specified and a value for the plate bending strength is required, then the terms "upper" and "lower" must be exchanged. For such reasons, the terms "safe" and "unsafe" are really more appropriate, though "upper" and "lower" are found in the literature more regularly than "unsafe" and "safe".

## 4.3 The normality rule

The yield criterion is a closed convex figure in moment space (that is moment coordinates). The purpose served by the criterion is to specify what combinations of bending moment, acting at a point in the plate, will cause yield. In addition to this moment information there is also some deformation information implied. This derives from the so-called *normality rule*.

Consider a point such as P in Fig. 4.2. The $M_2$ moment is at yield while the $M_1$ moment is not. The outward normal drawn to the yield surface at P points in a direction parallel to the $M_2$ axis. The normality rule then states that only curvature changes in this direction are possible at point P. In this case then the only possible curvature change is one associated with the yield moment $M_2 = M_p$ at P.

Consider Q (Fig. 4.2). Here, because of the role of the maximum stress (hence moment) difference noted earlier, the outward normal now points in a direction equally inclined to the moment axes. The interpretation now is that the increments of curvature change in the 2 and negative 1 directions must be equal at point Q.

Though very often the information provided by the normality rule will not be used fully, it nevertheless provides an important link between the moment information of the yield criterion and the deformation information on which the construction of mechanisms depends. Examples later will illustrate the use of the normality rule.

## 4.4 Circular plates—square yield locus

The most satisfactory way of illustrating the use of yield criteria, the bound theorems and the normality rule is by examples. To begin with consider the circular plate, subject to a uniform pressure loading, and suppose the plate to be clamped around the edge.

When thinking about plastic plate problems in particular, it is well worthwhile to visualize such things as the shape into which the plate might deflect. Usually the simplest possible shape, consistent with the type of support the plate must conform to, is the one to investigate, at least for a start on the problem. Thereafter related or more complicated shapes often are suggested by what has gone before. Most frequently, a work equation formulation of equilibrium will be used, rather than consideration of the differential equations of equilibrium. For this reason, too, the careful visualization of the deflected shape of the plate is very useful in formulating the relevant terms in the work equation.

In the present case of the circular plate, the most obvious mode of deformation to consider is one whereby the originally plane plate deforms into a pyramid shape. The simplest shape of this type is the triangular-based pyramid (Fig. 4.6*a*). Once $\Delta$, the central displacement, is specified, then the shape is determined. In particular, the angles formed between adjacent plane (deformed) faces of the deflected pyramid shape are determined. Thus the greatest slope on any pyramid face is $2\Delta/R$. But slope is a *vector* quantity, and hence the resolved part of this slope normal to the hinge line OA is

$$\frac{2\Delta}{R}.\operatorname{Cos}\pi/6 = \frac{\sqrt{3}\Delta}{R}. \tag{4.4.0}$$

Hence the total angle of rotation between the adjacent pyramid faces is $2\sqrt{3}\Delta/R$. This is the relative rotation across a radial hinge line when $n = 3$. Across an edge hinge line such as AB, the relative rotation is the earlier calculated maximum slope of $2\Delta/R$.

Consider now the material response. This is described by the yield criterion. The description "plastic" does not fully determine the material response, and the precise yield conditions must be spelled out. For simplicity here it will be assumed that the yield locus is as given in Fig. 4.1—this is the "square" yield locus of earlier and is a good approximation to certain physical material responses, notably reinforced concrete elements with equal reinforcement in each direction and in each face of the plate. But it is also a suitable simple yield condition for steel plates. Here a positive moment will be assumed hogging, and negative, sagging. $M_{1,2}$ refer to the *principal* moments at a given point in the plate. Thus from point to point, the *orientation* of these moments may vary. Such information is not explicitly

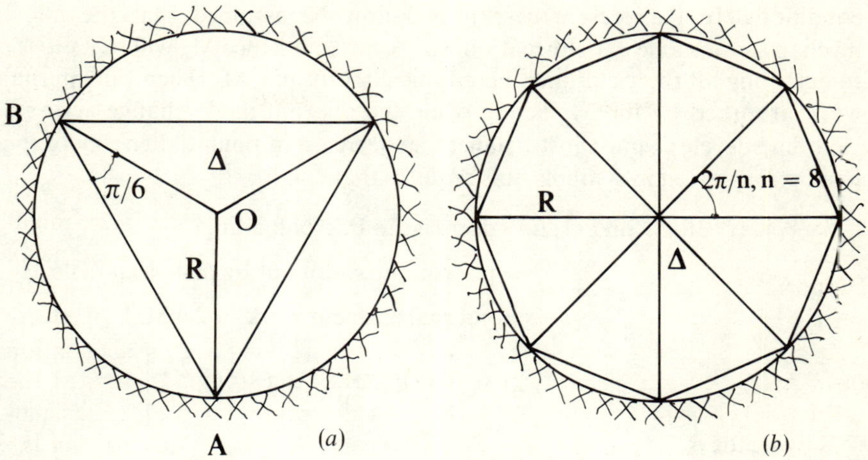


**Figure 4.6** Collapse of circular plate. (*a*) $n = 3$; (*b*) $n = n$.

provided by the yield locus. The non-dimensionalizing quantity, $M_p$, is the fully-plastic moment, analogous to the uniaxial beam fully plastic moment (see 4.1). In the present example, the $n = 3$ case, there will be moments in general generated throughout the plate when the pressure loading, p, rises to the collapse value, $p_c$.

To solve the problem completely, a solution for the internal moments must be constructed in which equilibrium is achieved at all points, yield is nowhere violated and yielding occurs at sufficient points for a yield mechanism to form. To satisfy all these conditions simultaneously is usually not possible. An approximation is sought, and the most fruitful approach is to invoke the bound theorems, and especially the upper bound (unsafe) theorem, assuming the collapse pressure ($p_c$) is being sought for a given strength ($M_p$) of plate. Now the search is relaxed to consider just the equilibrium and mechanism

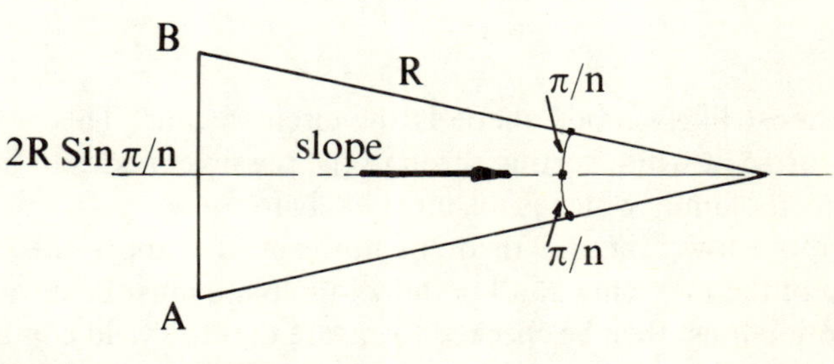


**Figure 4.7** Circular plate, typical segment.

conditions. In the present case it will now be assumed that the $n = 3$ mechanism or mode of deformation will occur, and hence $M_p$ will be *assumed* to act along all the yield lines, here typically OA and AB. Then the internal work absorbed by the $M_p$ acting while the relevant angle change across a yield line develops must just balance the decrease in potential energy of the load, p. Expressed in symbols, this requires that

$$p \times \tfrac{1}{3}.\tfrac{1}{2}.(\sqrt{3}R)^2.\operatorname{Sin}\pi/3.\Delta = \text{decrease in P.E. of loading}$$

$$= p \times \text{volume swept out by plate as it deflects}$$

$$= \text{internal moment work} = \Sigma|M_p|.|\theta|.L,$$

$$= M_p\left[3.R.\frac{2\sqrt{3}\Delta}{R} + 3\sqrt{3}R.\frac{2\Delta}{R}\right].$$

Solving for p,

$$p_3 = 48\frac{M_p}{R^2}. \tag{4.4.1}$$

This *estimate* of the true collapse load, $p_c$, is, according to the upper bound theorem, either equal to, or greater than, $p_c$. To investigate the solution further suppose the pyramid shape of the deformation is based on a regular "n" sided polygonal shape, rather than just the 3 sided shape.

Now, the greatest slope toward the centre is $\Delta/R\operatorname{Cos}\pi/n$ and this, resolved normal to the sagging (radial) yield line, is $\Delta/R\operatorname{Cos}\pi/n.\operatorname{Sin}\pi/n$. The work equation becomes

$$p.\frac{1}{3}n\left(\frac{1}{2}R^2\operatorname{Sin}\frac{2\pi}{n}\right) = M_p\left[n.\left(2R\operatorname{Sin}\frac{\pi}{n}\right).\frac{\Delta}{R\operatorname{Cos}\pi/n} + n.\frac{2\Delta}{R}.\operatorname{Tan}\frac{\pi}{n}.R\right]$$

or

$$p_n = \frac{12M_p}{R^2}\operatorname{Sec}^2\frac{\pi}{n}. \tag{4.4.2}$$

The minimum value of this expression is for $n = \infty$ when

$$p_c = \frac{12M_p}{R^2}. \tag{4.4.3}$$

Hence the most likely mode shape is the circular cone. This is the Upper Bound Theorem at work, sorting through the possible modes to find the one of the particular family which is the most likely to occur.

Consider the lower bound theorem however. To apply this, a detailed description of the moments at all points of the plate must be developed, and this description must then be checked to ensure that the yield condition is not exceeded (violated) anywhere, and also that equilibrium is satisfied.

Earlier, in Chapter 3, the equilibrium equations for the circular plate were established. These will now be used to develop a suitable description of the bending moments. From (3.3.3) the $M_{rr}$, $M_{\theta\theta}$ principal moments and transverse shear $Q_r$ are related by

and
$$(rM_{rr})_{,r} - M_{\theta\theta} - rQ_r = 0, \qquad (4.4.4)$$
$$(rQ_r)_{,r} = pr.$$

The transverse shear force, $Q_r$, can be eliminated to give

$$(rM_{rr})_{,rr} - M_{\theta\theta,r} = pr. \qquad (4.4.5)$$

The approach to using the lower bound theorem successfully is usually made more easy if some features of the *upper* bound solution, which would be sought first, can be incorporated. Here is it noted that the conical shape implied by the least upper bound is associated with a conical deflected shape and this in turn implies that there is yielding across radial lines, in order that the $\kappa_2$ curvatures can develop in the cone.

Hence it is concluded that it is likely that $M_{\theta\theta}$ will be at yield, i.e. $M_{\theta\theta} = -M_p$ at all points along a radius. When used in (4.4.5), this information gives a single equation for $M_{rr}$ from which by simple integration there is obtained

$$rM_{rr} = \frac{pr^3}{6} + Ar + B. \qquad (4.4.6)$$

The $M_{\theta\theta}$ term, note, is lost in (4.4.5) because it is a constant. Now to find the values for A and B in (4.4.6)—at r = O, $M_{rr} = M_{\theta\theta}$ from symmetry, and hence $M_{rr}(O) = -M_p$.

Thus

$$r = O, \quad O = O + O + B,$$

but then $M_{rr}(O) = O + A = -M_p$.

Hence

$$M_{rr} = \frac{pr^2}{6} - M_p. \qquad (4.4.7)$$

At the outer edge the radial moment is a maximum, but now hogging rather than sagging, and hence

$$M_{rr}(R) = +M_p.$$

Finally then

$$\frac{pR^2}{6} = 2M_p \quad \text{or} \quad p_L = \frac{12M_p}{R^2}. \qquad (4.4.8)$$

This $p_L$ is a *lower* bound estimate for the collapse pressure. It will be noted that it coincides in value with the least upper bound which has been constructed. The conclusion to be drawn is that since the upper and lower bounds coincide, so the value $p_c = 12M_p/R^2$ must be the *exact collapse load*. This example has been considered at length because of the importance of the solution being *exact*. In very few other instances will this prove possible.

Before leaving this example, consider the role of the Normality Rule. In the evaluation of the moments for the lower bound solution, the convention has been adopted that hogging moments are positive. Suppose we ask whether the moment combination of this analysis can be traced out on the yield locus to any purpose?

Consider the vertex O (Fig. 4.8). The coordinates of this point are those for the moments at the centre of the plate. Successive points along the lower edge of the Yield Locus are the coordinates of the moment components for successive points along a typical radius, with the vertex R being the combination of moments which occurs at the edge of the plate. Note, however, that according to (4.4.7) the $M_{rr}$ value varies parabolically and so the point $M_{rr} = 0$, $M_{\theta\theta} = -M_p$, namely point P, Fig. 4.8, corresponds to a point on the actual plate at radius r such that

$$0 = +\frac{p_c r^2}{6} - M_p, \quad \text{with} \quad p_c = \frac{12M_p}{R^2}.$$

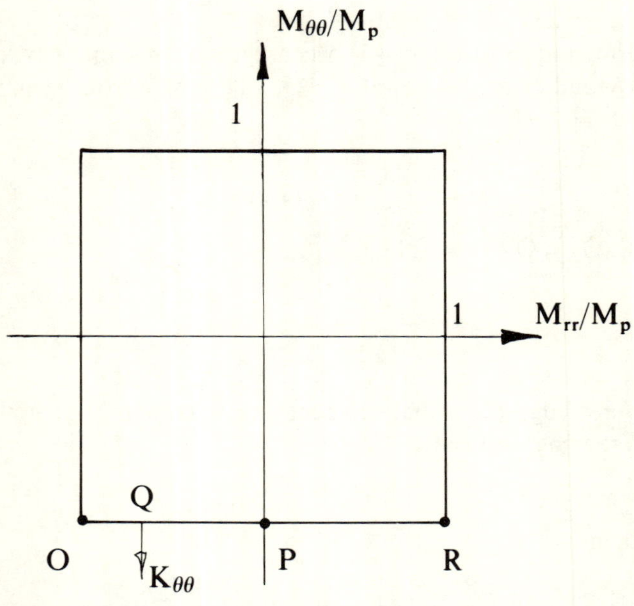


**Figure 4.8** Square yield criterion—polar coordinates.

Hence

$$r^2 = \frac{R^2}{2}$$

or

$$r = 0.707R. \qquad (4.4.9)$$

What can the normality rule contribute? Draw the normal to the yield locus for some typical point, such as Q (Fig. 4.8). Note that this normal points wholly in the direction of $-M_{\theta\theta}$, and, because of the shape of the yield locus, has no component in the $M_{rr}$ direction. The interpretation is therefore that the *allowable curvatures* according to the normality rule must be such that $\kappa_{\theta\theta}$ exists but $\kappa_{rr}$ is zero. The $\kappa_{\theta\theta}$ exists but the *value* is not determined. This is consistent with the appearance of an arbitrary $\Delta$, central displacement, in the upper bound solution. In effect any value of $\Delta$ and hence of relative $\kappa_{\theta\theta}$ values can occur, which corresponds to plasticity having set in and progressed to just such an extent as to have produced $\Delta$.

Now, referring to the earlier discussion of surface curvature, the requirement that $\kappa_{\theta\theta}$ be present but $\kappa_{rr}$ be zero can be seen to be consistent with the *conical* shape which has been produced as most likely by the upper bound calculations. It is therefore seen that the link between the two bounds has now been established quite firmly, since the lower bound use of the yield locus has suggested the use of the normality rule and this in turn has suggested the conical deformation mode, which was quite independently produced by systematic application of the upper bound theorem.

*Exercise:* Consider a simply supported circular plate, radius R, pressure loaded. Show that the exact solution for $p_c$ is given by

$$\frac{p_c R^2}{M_p} = 6.$$

Confirm that the mode of collapse is conical and that the moments are everywhere sagging.

## 4.5 Circular plates—Tresca yield locus

In this section the problem considered in 4.4 will be reworked, the only change being the adoption of the Tresca rather than the square yield criterion (see Fig. 4.9). Differences will be apparent as the solution is developed, and clearly these will arise from the associated yet different yield criterion being used. First note that the square criterion circumscribes the Tresca locus. From this observation it can be concluded that the upper bound calculation in 4.4 can be carried directly over to this problem and will still represent an upper bound, but probably a less good one. That is to say,

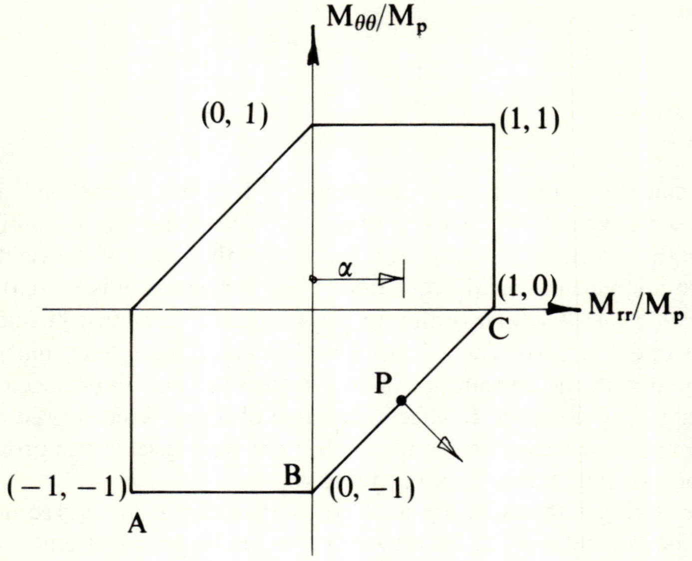


**Figure 4.9** Tresca yield criterion—polar coordinates.

since the yield locus essentially implied by the upper bound calculation of 4.4 assumes $M = M_p$ along all yield lines, this may be an overestimate of the available yield capacity and hence may suggest a higher carrying capacity for the plate than is available when the material responds according to the Tresca locus. This will be particularly true in regions of the plate where the principal ($M_{rr}$, $M_{\theta\theta}$ ) moments are of opposite sign, namely near the outer edge.

What then can be said about the lower bound calculation? The solution can begin much as before by noting that at the centre of the plate the moment field will be sagging and at the centre the moments can be thought of as described by vertex A, Fig. 4.9. Indeed, the plate inside some radius $\rho < R$, to be determined such that $M_{rr}(\rho) = 0$, behaves exactly as does a similar plate responding according to the square yield locus. It is expected that a conical mode shape will develop for $0 < r < \rho$, and the moments will be given by

$$M_{\theta\theta} = -M_p, \quad M_{rr} = \frac{pr^2}{6} - M_p. \tag{4.5.0}$$

Hence $\rho$ is given by $M_{rr}(\rho) = 0$ or

$$\rho^2 = \frac{6M_p}{p}. \tag{4.5.1}$$

Recall that p is presently an unknown: indeed the primary object is to evaluate p, either as $p_L$ or $p_c$, the lower bound pressure or the true collapse pressure, should it prove possible to satisfy all requirements of the problem. Thus $\rho$ will not be 0.707R as in the previous problem where $p = 12M_p/R^2$; p is now expected to be *less* than this value because the yield locus can be *inscribed* in the previous yield locus.

For radii $\rho \leqslant r \leqslant R$ a new solution must be sought to the equilibrium equations which incorporates accurately the requirements of the Tresca Yield Locus. Now $r = \rho$ corresponds to vertex B on the Yield Locus. As r extends beyond $\rho$ it is tentatively proposed, to be checked later, that the moment combinations correspond to successive points along the segment BC of the yield locus. If this is so then it must be the case that for

$$\rho \leqslant r \leqslant R, M_{rr} - M_{\theta\theta} = M_p,$$

since this is the equation of the face BC of the yield locus. But equilibrium still demands that $rQ_r = pr^2/2 + A$ and $A = 0$ since there are no point loads present.

Thus

$$(rM_{rr}),_r - M_{\theta\theta} = \frac{pr^2}{2}$$

and hence

$$(rM_{rr}),_r - (M_{rr} - M_p) = \frac{pr^2}{2}.$$

There is a cancellation of $M_{rr}$ terms and on cancelling by r then integrating, we obtain

$$M_{rr} = \frac{pr^2}{4} - M_p \ln r + \alpha. \tag{4.5.2}$$

The region interior to $r = \rho$ must match with that exterior to $r = \rho$, at least in respect of the values of $M_{rr}$ and $Q_r$. Hence

$$M_{rr}(\rho) = \frac{p\rho^2}{4} - M_p . \ln r + \alpha = 0 \quad (M_{\theta\theta} = -M_p) \tag{4.5.3}$$

and

$$Q_r(\rho) = \frac{p\rho}{2}$$

from either side of $r = \rho$.

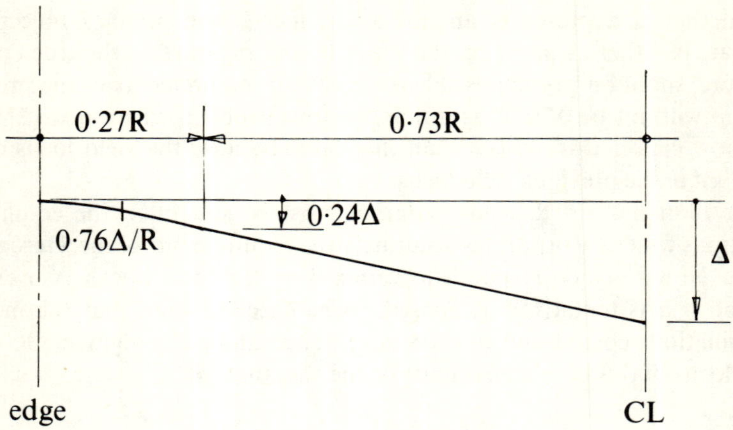


**Figure 4.10** Tresca plate—section at collapse.

The first of (4.5.3) is an equation for $\alpha$ in terms of p, $\rho$ and $M_p$. Thus

$$-\alpha = \frac{p\rho^2}{4} - M_p \ln \rho = 1.5M_p - M_p \ln \rho \qquad (4.5.4)$$

after (4.5.1) has been used.

But at r = R, the outer edge, suppose the $M_{rr}$ value has reached some value associated with a point such as P (Fig. 4.10) with coordinates $(\alpha, -1+\alpha)$.

Then

$$\alpha M_p = \frac{pR^2}{4} - M_p . \ln R - (1.5 - \ln \rho)M_p,$$

$$= M_p\left[1.5\left(\frac{R}{\rho}\right)^2 - \ln\frac{R}{\rho} - 1.5\right]. \qquad (4.5.5)$$

In the spirit of a lower bound solution, the aim is to maximize p, i.e., minimize $\rho$. Expressed alternatively, (4.5.5) is an equation for $R/\rho$ and the aim is to maximize this value. This is achieved when $\alpha$ is at its maximum of 1. Then

$$1.5\left(\frac{R}{\rho}\right)^2 - \ln\frac{R}{\rho} = 2.5$$

or $R/\rho = 1.37$, $\rho/R = 0.73$, and hence, from (4.5.1)

$$p_L = \frac{6M_p}{\rho^2} = \frac{6M_p(1.37)^2}{R^2} = 11.25\frac{M_p}{R^2}. \qquad (4.5.6)$$

As observed earlier, this lower bound value is less than that associated with the square yield locus example for the reason that the material is weaker if responding according to the Tresca-type yield behaviour, for the same value of $M_p$.

Of perhaps more interest than the $p_L$ value is the use of the normality rule to investigate the *shape* into which the plate is deformed in this case.

First note that there is a central conical region described by the normal to the yield locus having no component in the $\kappa_{rr}$ direction and hence $\kappa_{rr} = 0$.

Now from earlier this is seen to require that

$$w_{,rr} = 0, \quad \text{or} \quad w = \alpha r + \beta. \tag{4.5.7}$$

In the outer region, for $r \geqslant \rho = 0.73R$, the normal to the yield surface between B and C is now seen to be equally inclined to the principal moment axes. More precisely, this normal, taking account of signs, is such that $\kappa_{\theta\theta}$ and $-\kappa_{rr}$ changes must be equal.

Hence

$$\kappa_{\theta\theta} = -\kappa_{rr}.$$

Expressed in terms of the transverse displacement and when the shallowness and surface of revolution properties are included, then

$$\frac{1}{r} w_{,r} = -w_{,rr}$$

or

$$w_{,rr} + \frac{1}{r} w_{,r} = 0. \tag{4.5.8}$$

This is a differential equation for w which can be solved as

$$(rw_{,r})_{,r} = 0 \quad \text{or} \quad rw_{,r} = A$$

Hence

$$w = A \ln r + B. \tag{4.5.9}$$

The constants A and B can be found, in terms of the overall central displacement of the plate, $\Delta$, by matching the displacement and slope of the plate at $r = \rho$ and noting that $w(R) = 0$.

Thus from (4.5.7, 8)

$$\beta = \Delta, \quad A \ln R + B = 0,$$

$$A \ln \rho + B = \alpha\rho + \Delta$$

and

$$\frac{A}{\rho} = \alpha . (\rho = 0.73R).$$

From these four equations

$$A \ln 1.37 = -A - \Delta \quad \text{or} \quad (1 + 0.315)A = -\Delta.$$

Hence, $A = -0.76\Delta$, $\alpha = -1.045/R$, $B = A \ln R = 0.760\Delta . \ln R$.
Displacement at $r = \rho$

$$w(\rho) = -0.760\Delta \ln \rho + 0.760\Delta \ln R$$

$$= 0.760\Delta \ln \frac{R}{\rho}$$

$$= 0.24\Delta$$

and hence the section on the deflected plate middle surface is as shown in Fig. 4.10.

To summarize, the value of $p_L$ (4.5.6) has now been shown to be $p_c$, the collapse load, since a mode of deformation has been associated with the lower bound solution. If the solutions derived in (4.4) and (4.5) are compared, the effect of a change in yield criterion can be seen to lower the collapse load because the Tresca Yield criterion can be inscribed inside the other (square) criterion. The deformed shape changes in the outer zone from conical (square locus) to a "hat-brim" sort of shape (Tresca), the result of the inclined face BC on the yield criterion, Fig. 4.9.

These two sections (4.4, 4.5) are important because they describe two of the few known exact solutions to plastic plate problems. In another sense they are important too—the circular plate represents a yardstick because the loaded area is enclosed by a minimum length boundary. Perhaps of more importance, the circular plate under pressure loading, and (say) a simply supported edge, is the *weakest* plate of given plan area (A) and given section strength ($M_p$). Expressed another way, the quantity $p_c A/M_p$ takes on a *minimum* value for the circular plate. It is not intended to prove this result, but intuitive arguments can be put forward to support the observation. The consequence of any such discussion is to further focus attention on the importance of the circular plate solution.

*Exercise:* Confirm that the solution for the simply supported circular plate, pressure loaded and subject to the Tresca yield criterion is identical to that for the square yield criterion.

## 4.6 Plates of other shapes—square and regular shapes

The circular plate has been discussed in some detail, because this problem will form a standard against which others can be compared. In this section other shapes will be considered, though in less detail, and mostly only as upper bounds on collapse load. The shape of most interest here is the rectangle.

Throughout this section it will be assumed that the plates are made from material which is isotropic and yields according to the *square* yield locus in principal moment space. This choice ensures minimum analytic complication while retaining all the essential features of collapse load calculation.

The bound theorems will be used, whenever appropriate, though the upper bound calculation will be seen to be much easier to perform. In a minority of practical examples only will it be found feasible to construct useful lower bound solutions for the collapse pressure. Consider then the problem of a square, uniform, simply supported plate, side length L, loaded with a uniform pressure (p). The object is to calculate the collapse value of p.

Begin with an upper bound calculation. The simplest mechanism is a system of diagonal yield lines radiating into the plate corners. Suppose the central deflection is an arbitrary amount, $\Delta$. If the plate is deforming plastically, then the calculation about to be made is an equilibrium one, which balances the work absorbed in the assumed hinge lines by the internal moments with the load work given up as the central deflection *increases* (from zero) by some arbitrary amount, $\Delta$.

Thus the (small) angle $\alpha$ developed across a yield line is $\alpha = 2\sqrt{2}\Delta/L$ and the moment work will be the product of the yield moment (m)/unit length, times the length of line, times the rotation angle developed. Note that the angle $\alpha$ is constant along each line. This amount of work absorbed is supplied by the decrease in potential energy of the load (p), and this is the area integral of p $\times$ local transverse displacement. This reduces to p $\times$ volume swept out by the plate, as it deflects from the originally plane shape to the final pyramid shape.

Hence

$$4 . m . \frac{2\sqrt{2}\Delta}{L} . \frac{L}{\sqrt{2}} = \frac{1}{3} . p . L^2\Delta$$

or

$$\frac{p_1 L^2}{m} = 24. \tag{4.6.0}$$

This is an upper bound estimate of $p_c$, the collapse pressure. Hence

$$p_1 \geqslant p_c.$$

In order to gauge by how much $p_1$ is in excess of the exact $p_c$ it is necessary to assume some more comprehensive mechanism which contains some parameters describing the deflected shape, then choose values for the parameters which will give a minimum value for p associated with this mechanism. These procedures will be illustrated in later sections. Here it is proposed to attempt the construction of an equilibrium moment field which can be used in constructing a lower bound estimate of $p_c$.

Choose an origin at the plate centre with axes x, y parallel to the plate edges. Then consider the bending moment distributions

$$M_{xx} = -\frac{pL^2}{24} + \frac{px^2}{6}$$

$$M_{xy} = +\frac{pxy}{6} \tag{4.6.1}$$

$$M_{yy} = -\frac{pL^2}{24} + \frac{py^2}{6}.$$

It is not obvious where these expressions have been derived from, but consider the form of the expressions. First along the edges $x = \pm L/2$, $M_{xx} = 0$ as required by the simple support boundary conditions. Indeed all the moment boundary conditions are met. Further, these expressions (4.6.1) can be shown, by direct substitution, to satisfy the plate equation (2.4.3). The loading at any point is carried equally by the bending in two directions and the twisting moments.

Consider further the values of the principal moments and where the largest principal values might be. For example, at the origin $M_{xx} = M_{yy} = -pL^2/24$, $M_{xy} = 0$. At a corner $M_{xx} = M_{yy} = 0$, $M_{xy} = pL^2/24$. If this system of moments is examined further it will be found that along any radial line the larger principal moment occurs for bending *about* the radial line and has a value equal to $pL^2/24$. Hence if

$$\frac{p_L L^2}{24} = M_p \tag{4.6.2}$$

then the yield condition is not violated anywhere. Indeed it is *met* in at least one direction *everywhere*. This $p_L$ is a lower bound and will be seen to be equal to the upper bound value $p_u$ (4.6.0). The conclusion to be drawn is that the *exact* solution $p_c$ is also given by the same value

$$\frac{p_c L^2}{M_p} = 24. \tag{4.6.3}$$

Hence the square, simply supported pressure loaded plate of side L will collapse under this value of pressure, if the *square* yield locus applies. Indeed it can be shown that any *regular* n-sided ($n \geqslant 3$) *simply supported plate* has an exact solution for $p_c$ given by

$$\frac{p_c R^2}{M_p} = 6\sec^2\frac{\pi}{n} \tag{4.6.4}$$

where R is the *radius* of the circumscribing circle.

*Exercises:* (1) The exploration of formula (4.6.4) is left as an exercise for the reader. Note that (4.6.4) includes (4.6.3) and the exercise in 4.4 as special cases.

(2) Repeat the upper bound calculations for collapse load on the square clamped plate *under pressure loading* and show that

$$\frac{p_u L^2}{M_p} = 48.$$

Now no simple lower bound solution is available (but see chapter 6).

(3) The circular plate, either simply supported or clamped, and pressure loaded, represents a very important standard against which other shapes can be compared. Confine attention to plates which obey the square yield criterion. Then explore the proposition that the parameter $\gamma = p_c A/M_p$ takes on a *minimum* value for the circular plate. Here $p_c$ is the collapse pressure, A is the plan area of the plate, or portion of it, which deforms, and $M_p$ is the full plastic moment/unit length of the isotropic plate.

If the plate is simply supported then the minimum value is $6\pi$, if clamped the minimum value is $12\pi$. Typically, the solution for the square, simply supported plate has already been shown to produce a $\gamma$ value of 24, which is greater than $6\pi$.

Especially, consider some shapes which lack total symmetry.

## B. REINFORCED CONCRETE PLATES

### 4.7 Yield-line theory—I. Fundamentals

Thus far the theory and examples presented for plastic plates have been concerned with the homogeneous type most typically represented by solid metal plates. In this section the interest is in reinforced concrete plates of the type often encountered for example in floor systems in buildings.

Though there are many points of similarity in method between metal and concrete plates, there are also important differences, which are the justification for a separate treatment, and the most important of these is in the material response to bending moment.

The physical situation is that the resistance of a concrete plate to bending resides in the reinforcement provided. Reinforcement parallel to the x axis provides resistance principally in the $M_{xx}$ direction, that is about an axis normal to the x direction. Resolved on to an axis which makes an angle of $\theta$ with the $M_{xx}$ axis, this moment of resistance provides a moment of $M_{xx} \cdot \mathrm{Cos}^2\,\theta$. Thus the moment of resistance resolves as a second order (tensor) quantity and not as a vector (first order tensor) (recall that vector quantities resolve as $\mathrm{Cos}\,\theta$ and not $\mathrm{Cos}^2\,\theta$).

If more than one system of reinforcing bars is incorporated, then the resistance moment across any chosen direction is taken to be the *summation* of the resistance moments provided in this direction according to the expression just quoted. If the direction is not principal then twisting moments will also act. In general twisting moments however play little part in the mechanics of concrete plates at ultimate load.

Consider an orthogonal system of reinforcement with moment of resistance $m_1$ in the x direction and $m_2$ in the y direction, for sagging bending moment. Then the *axis* of the $m_1$ moment of resistance will be parallel to the y axis; that is, normal to the reinforcement direction (Fig. 4.11). Likewise, for $m_2$ the axis for bending is parallel to the x axis. Suppose for the present that $m_1 \neq m_2$; then the system is described as *orthotropic*.

For a line making an angle $\theta$ with the x axis (Fig. 4.11) the combined moment strength *about* this line of the two systems of reinforcement is given by

$$\begin{aligned} m_\theta &= m_2 \operatorname{Cos}^2 \theta + m_1 \operatorname{Cos}^2 \left(\frac{\pi}{2} - \theta\right), \\ &= m_2 \operatorname{Cos}^2 \theta + m_1 \operatorname{Sin}^2 \theta. \end{aligned} \tag{4.7.0}$$

In the special case when $m_1 = m_2$, then this orthogonal arrangement of reinforcement is called *isotropic* and (4.7.0) shows that

$$m_\theta = m_1 \operatorname{Cos}^2 \theta + m_1 \operatorname{Sin}^2 \theta = m_1 \equiv M_p, \tag{4.7.1}$$

namely the strength of the slab is equal in all directions. Isotropic and orthotropic systems of reinforcement are the two most important systems encountered in practical applications.

The isotropic system described above can conveniently be represented in yield condition terms by the square yield criterion of earlier (4.4, Fig. 4.1) on axes of principal moment—the $m_1$, or $m_2$ $(=m_1)$ axes. This yield criterion states that each of $m_1$ and $m_2$ can rise to $M_p$, the full plastic value, independently, and the yield condition is just satisfied. In the conventional study of concrete slabs by the yield line method, explicit reference to the yield locus (or criterion) is seldom made because the calculation made is an upper bound on the collapse load, and in this attention is focused on the mechanism and equilibrium conditions. The yield condition is *assumed* to be met across all of the assumed yield lines, but this is not rigorously ensured, so the calculation can only be expected to be an upper bound on the collapse load.

The concept of a *mechanism of failure* has already been ençountered (4.3, 4.4). This concept will now be extended and used in conjunction with a work calculation as the main solution process for slab problems by the *yield line method.*

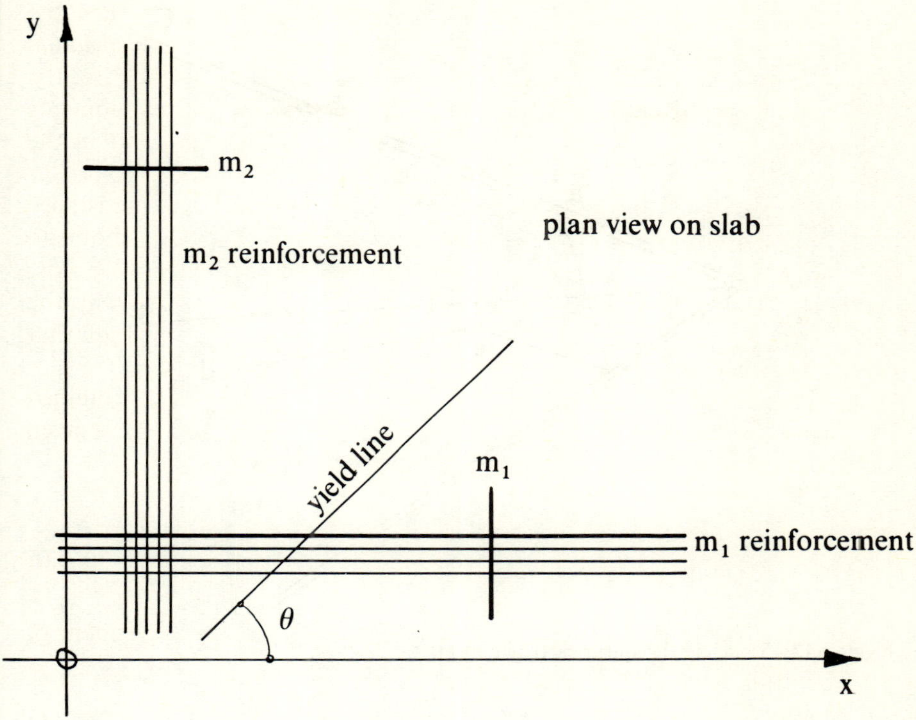


**Figure 4.11** Reinforcement and moment capacities,

A mechanism is a combination of directions in the slab about each of which rotation is assumed to occur, so producing movement from the original plane slab to some shallow pyramid-shaped surface. All the directions which will be considered here will be straight. This is adopted for simplicity and means that all the deflected surfaces will have plane faces which meet other plane faces along the yield lines. To be a possible mechanism there are certain requirements of the yield line arrangement, but it is not easy to set these down in the form of adequate rules.

Consider the corner of a slab (Fig. 4.12). If the slab in the corner is to deform by yield line rotation then rotation must also occur about the simply supported edges. Indeed these edges can be thought of as yield lines of zero strength. Hence the sagging yield line AB must radiate *into* the corner A, and rotation of leaf I of slab with respect to leaf II about AB is accompanied by rotation about AC and AD, allowing B to descend and relative rotation about AB to occur.

But it is possible for other arrangements of yield lines into the corner to occur, for example, the hogging line ab associated with the sagging lines ac,

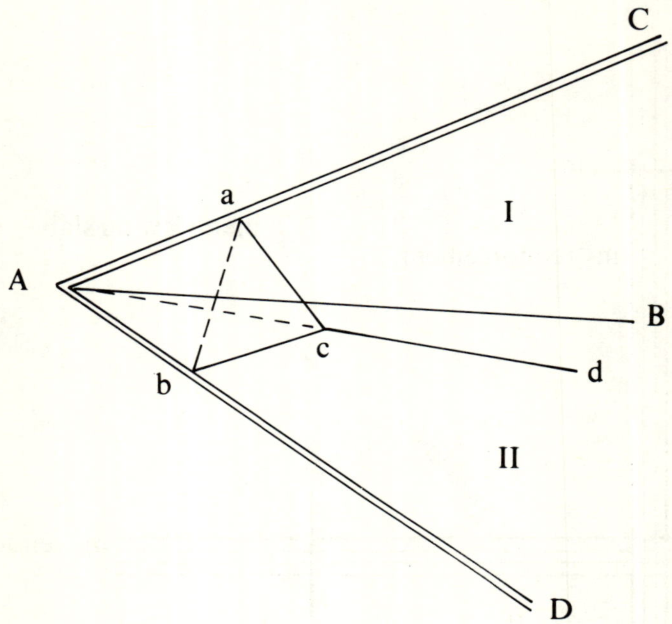


**Figure 4.12** Yield lines—geometrical requirements.

bc, cd. This is a valid system provided that dc (produced) passes through A, for unless this is so the rigid leaves of slab spanning on to the supports at aC and bD could not rotate. In this system of yield lines the hogging line (ab) has in effect become part of the boundary, and would be a line of zero moment strength if no hogging strength was provided.

Specifying the transverse displacement of d fixes the amounts of rotation of the slab portions about aC and bD and hence too the displacement of c. Then the rotation of abc about ab is also fixed. Hence all the movements and rotations relate back to that of d. This is spoken of as a *one degree of freedom* system. These are the most usual types of mechanism, though multi-degree of freedom systems are possible.

Yield line patterns may be specified by *parameters* and these parameters should be chosen to give as low an estimate of p as possible. For example a, b and c can be varied in position. But as a piece of practical advice it is suggested that as *few parameters as possible* be associated with a given mechanism—at least initially.

Consider the class of rectangular simply supported slab shapes (Fig. 4.13*a*).

Then the crudest mechanism possible consists of two intersecting yield lines which run from one corner to an opposite corner. If it is assumed that

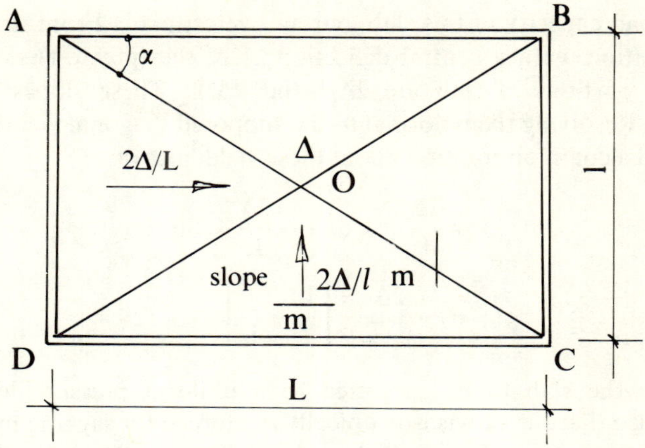


**Figure 4.13**(*a*) Example—rectangular plate I.

the slab has no hogging resistance, then there is no restraint to the edges of the slab rotating about their supports. Suppose sagging yield occurs along these two diagonal yield lines to give a maximum transverse deflection at the centre point of Δ (say). Here Δ is the one degree of freedom. Then by rotation about the edge lines of the four respective portions of slab deflect into the shallow pyramid shape with maximum central deflection of Δ, (Fig. 4.13*c*).

In reality, if an experiment was to be conducted this might not be the actual collapse mechanism to develop. However, it is likely that an Upper Bound calculation based on this mechanism will be a reasonable estimate of the

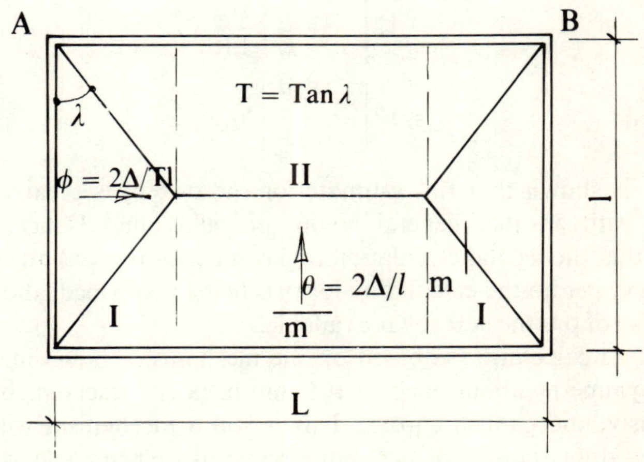


**Figure 4.13**(*b*) Example—rectangular plate II.

collapse load capacity of this slab, but an overestimate. From the geometry of deformation, with a central deflection of $\Delta$, the (greatest) slopes of the respective portions of slab are $2\Delta/l$ and $2\Delta/L$. These slopes are vector quantities. Resolving them normal to the supposed diagonal yield lines, there is obtained a common rotation about these yield lines of

$$= \frac{2\Delta}{l} \times \operatorname{Cos}\alpha + \frac{2\Delta}{L} \times \operatorname{Sin}\alpha,$$

$$= \frac{2\Delta}{\sqrt{L^2+l^2}}\left[\frac{L}{l}+\frac{l}{L}\right].$$

Suppose the slab to be subjected to a uniform pressure loading, $p_1$. Suppose also that the slab is isotropically reinforced for sagging moments, to a strength $m_1 = m_2 = M_p$ parallel to the slab edges. Then, by the analysis earlier in this section, the strength across any possible sagging yield line will also be $M_p$. Use a work calculation in which a virtual central displacement of $\Delta$ associated with the above mechanism produces internal work absorption along the yield lines; this is to be balanced by loss of potential energy of the loading. Then the virtual work balance will appear as

$$\frac{2\Delta}{\sqrt{L^2+l^2}}\left[\frac{L}{l}+\frac{l}{L}\right].\, M_p \times 2\sqrt{L^2+l^2} = \text{internal moment work absorbed}$$

$$= \text{load work done} = \frac{p}{3}\Delta L l = (p \times \text{volume of pyramid shape}).$$

Hence, solving for $p_1$, and noting that the virtual $\Delta$ cancels, obtain

$$p_1 = 12\left[\frac{L}{l}+\frac{l}{L}\right]\frac{M_p}{Ll}$$

$$= 12\left[\frac{L^2+l^2}{(Ll)^2}\right]M_p. \qquad (4.7.2)$$

Below it is shown that this estimate for the true $p_c$ is greater than that associated with another general layout of yield lines. Hence, it can be concluded that the second calculation ($p_2$) is superior to $p_1$ as an estimate for $p_c$, since an upper bound calculation for $p_c$ is being performed, and the lowest current value of p is the *best* so far available.

This second calculation is based on the mechanism shown in Fig. 4.13*b*. Here the symmetry about each of the mid-lines is preserved, but a more comprehensive mechanism chosen. This is still a mechanism with just *one* degree of freedom, thought of as $\Delta$, but $\lambda$ is an unspecified angle, which in the previous mechanism was fixed as $T = \operatorname{Tan}\lambda = L/l$.

Leaving $\lambda$ as a parameter, the external load work and moment work can be evaluated as previously to give

$$\text{External load work} = \frac{1}{3}(l\,.\,l\mathrm{T})\mathrm{p}_2\Delta + (\mathrm{L}-l\mathrm{T})\frac{1}{2}\cdot \mathrm{p}_2\Delta$$

$$= \mathrm{M_p}\left[2l\,.\,\frac{2\Delta}{\mathrm{T}l} + 2\mathrm{L}\,.\,\frac{2\Delta}{l}\right] = \text{internal moment work.}$$

Here the evaluation has been arranged to split the load work so that the two zones I contribute equally to the first load work term, where zone II accounts for the second. The internal moment work can alternatively be derived exactly as in the previous calculation but as an angle of rotation of $\Phi_1$ across the diagonal lines of

$$\Phi_1 = \frac{2\Delta}{\mathrm{L}}\,.\,\mathrm{Sin}\,\lambda + \frac{2\Delta}{\mathrm{T}l}\,.\,\mathrm{Cos}\,\lambda$$

acting on moment $\mathrm{M_p}$ along a yield line of length $4 \times \mathrm{AO}$, together with $\Phi_2 = 4\Delta/\mathrm{L}$ acting on $\mathrm{M_p}$ along the central yield line of length $(\mathrm{L}-l\mathrm{T})$.

Simplifying (4.7.3) and solving for $\mathrm{p}_2$ there is obtained an expression still containing the parameter $\mathrm{T} = \mathrm{Tan}\,\lambda$. As $\lambda$ is varied so $\mathrm{p}_2$ will change. What is sought is the value of $\lambda$ for which $\mathrm{p}_2$ takes on a minimum value. Now

$$4\mathrm{M_p}\left[\frac{1}{\mathrm{T}} + \frac{\mathrm{L}}{l}\right] = \frac{\mathrm{p}_2 l^2}{6}\left[\frac{3\mathrm{L}}{l} - \mathrm{T}\right]. \qquad (4.7.3)$$

This minimum for $\mathrm{p}_2$ can most easily be sought by differentiating $\mathrm{p}_2$, regarded as a function of T, with respect to T, rather than $\lambda$. There is then obtained a lengthy expression, since $\mathrm{p}_2$ is a quotient in T. The minimum is associated with a value of zero for the *numerator* in this expression.

Hence

$$\left[\frac{1}{\mathrm{T}} + \frac{\mathrm{L}}{l}\right](-1) = \left[\frac{3\mathrm{L}}{l} - \mathrm{T}\right]\left(-\frac{1}{\mathrm{T}^2}\right).$$

Simplifying, we obtain

$$\left(\frac{\mathrm{L}}{l}\right)\mathrm{T}^2 + 2\mathrm{T} - \left(\frac{3\mathrm{L}}{l}\right) = 0.$$

This quadratic in T provides only one physically interesting root, namely

$$\mathrm{T} = \frac{\sqrt{3(\mathrm{L}/l)^2+1}-1}{(\mathrm{L}/l)}. \qquad (4.7.4)$$

Then, typically, for the case $L/l = 1$, $T = 1$ and

$$p_2 = 24\frac{m}{l^2}.$$

This is the same value as given by $p_1$ in (4.7.2).

But this should be so since for $L/l = 1$, since then the two mechanisms are identical.

However for $L/l > 1$, $p_2 < p_1$ as shown by the following table, indicating that $p_2$ is the superior value.

**Table 4.0** Collapse pressures for two mechanisms

| L/l | $k_1$ | $k_2$ |
|---|---|---|
| | $p_1 = k_1\frac{M_p}{l^2}$ (Fig. 4.13*a*) | $p_2 = k_2\frac{M_p}{l^2}$ (Fig. 4.13*b*) |
| 1 | 24 | 24 |
| 2 | 15 | 14.1 |
| 3 | 13.3 | 11.7 |
| α | 12 | 8 |

The $p_2$ values have been found as the result of a minimization, involving differentiation and solution of a quadratic equation—and this with just a single parameter in the mechanism choice. In many practical problems it is not either feasible or worthwhile, in terms of the final result, to contemplate such a procedure.

Instead some reasonable choice of value for the free parameter, or some few reasonable choices, followed by an *evaluation* of the p expression should suffice to give an adequate exploration of the chosen mechanism. In respect of the example discussed above, a reasonable choice for T might be unity together with say 0.75 and 1.5, to give three values for $p_2$ and indicate the effect on $p_2$ of changes in T. By plotting $p_2/T$, the minimum or near minimum value can then usually be detected.

In the example just discussed an implicit use of the so-called *projected lengths* formula has been made. Consider a piece at yield line ab between two pieces of slab, and consider the edges about which these pieces of slab I, II are rotating (Fig. 4.14). For ab to be a possible yield then ab extended must pass through the point where the two edges meet. Suppose the angles $\alpha$, $\beta$ define the position of the yield line with respect to these edges (Fig. 4.14). Then if $\theta$, $\phi$ are the maximum slopes of the slab portions I, II, the angle between I, II across ab is given by $\theta . \text{Cos}\,\alpha + \phi . \text{Cos}\,\beta$, and the moment work along this line ab is $M_p(ab)\,(\theta\,\text{Cos}\,\alpha + \phi\,\text{Cos}\,\beta)$.

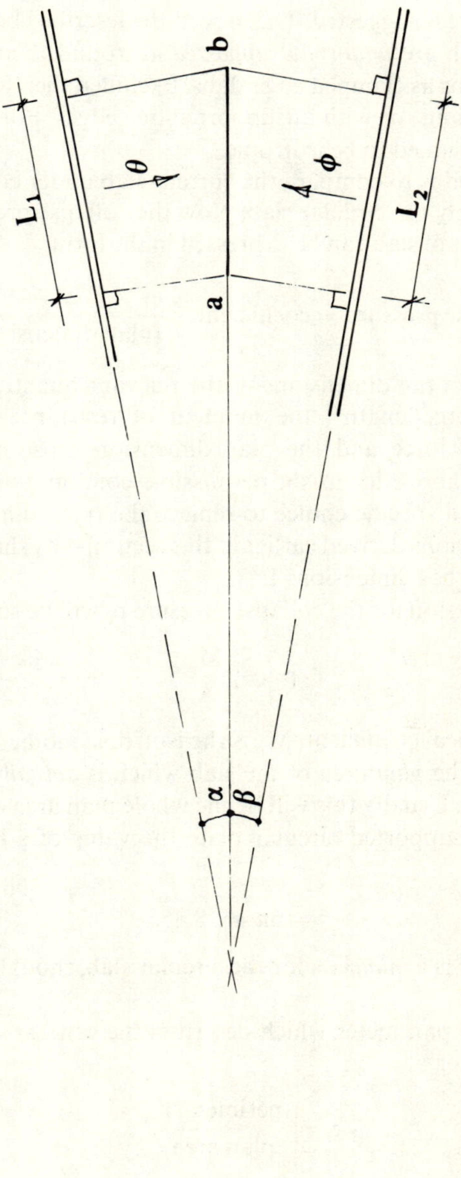


**Figure 4.14** Projected lengths of a yield line.

Rearrange this expression to read $M_p(\theta L_1 + \phi L_2)$, then this is the *projected lengths* formula for the piece of yield line ab. Here $L_1$, $L_2$ are the projections of ab on the two edges respectively. The technique then is to project the yield line onto the edges and sum the products of projected length $\times$ maximum slope.

As an aid to deciding whether the current estimate of $p_u$ is satisfactory, the following procedure is suggested. The procedure described below is applicable only to slabs which are *uniformly supported* all round their edges, either as simply supported, or as clamped edge slabs. Excluded therefore are slabs with mixed edge conditions, or with unsupported free edges. For the present, too, the slabs will be assumed to be isotropic.

What is proposed is to compare the current slab under consideration with an *ideal* slab, namely the circular slab. Now the collapse pressure for a given strength and layout of slab can be expressed in the form

$$\text{collapse pressure} = \text{coefficient} \times \frac{M_p}{(\text{plan dimension})^2}.$$

This follows from the dimensions of the relevant quantities; the pressure has dimensions force/length$^2$, the moment of resistance per unit length has dimensions of force and the plan dimension those of length. Hence dimensionally the above form is a permissible combination. However, it is proposed to make a specific choice to replace the (plan dimension)$^2$ term—the $l^2$ of the expressions derived earlier in this section—by the choice *slab plan area*, A, which also has dimensions $L^2$.

Hence the expression for the collapse pressure $p_c$ will be sought in the form

$$p_c = \gamma \frac{M_p}{A} \tag{4.7.5}$$

where $\gamma$ is a numerical coefficient, $M_p$ is the isotropic moment of resistance of the slab and A is the *plan area* of the slab which is *actually spanned* by the chosen mechanism. Usually this will be the whole plan area of the slab.

For the simply supported circular plate the value of $\gamma$ has already been shown (4.4) to be

$$\gamma = 6\pi = 18.85. \tag{4.7.6}$$

This coefficient $\gamma$ is a *minimum* for the circular slab, though no proof of this will be offered here.

There is a useful parameter which describes the general *shape* of the slab. This is the parameter

$$\alpha = \frac{(\text{perimeter})^2}{\text{plan area}}. \tag{4.7.7}$$

Note that both $\alpha$ and $\gamma$ are *pure numbers*. Now $\alpha$ also has a *minimum* value for the circular slab of $(2\pi r)^2/\pi r^2 = 4\pi = 12.57$. All other slab shapes have values of $\alpha$ in excess of $4\pi$.

Consider finally a third non-dimensional parameter $\beta$ given by

$$\beta \equiv \frac{\gamma}{\alpha}. \tag{4.7.8}$$

Then for the circular simply supported slab

$$\beta = \frac{6\pi}{4\pi} = 1.5. \tag{4.7.9}$$

It is an observation which can be tested with examples that $\beta$, evaluated for many simply supported isotropic plates under pressure loading, *is greater than or equal to 1.5*, and is usually in the range 1.5–1.6. Should it happen that the value of $\beta$ is greater than about 1.6 then the slab is elongated in one direction and hence can probably be treated as almost a one-way spanning system.

This observation can be used in a variety of ways. First, either the collapse pressure ($p_c$) or the required slab strength (m) can be solved for if the other quantities are known. Alternatively, computing the value of $\beta$ at the conclusion of an independent calculation for $p_c$ or $M_p$ can serve as a quick and useful check on the correctness, or at least the plausibility, of the result.

For example, the rectangular slab discussed earlier in this section produces the $\beta$ values in Table 4.1.

**Table 4.1** Rectangular slab — non-dimensional parameters (see Fig. 4.13*b*)

| L/l | $\gamma$ | $\alpha$ | $\beta$ |
|---|---|---|---|
| 1 | 24 | 16 | 1.50 |
| 2 | 28.3 | 18 | 1.57 |
| 3 | 35.2 | 21.3 | 1.65 |
| 4 | 42.7 | 25 | 1.71 |

*Exercise:* The discussion thus far has dealt with simply supported slabs. Show that if the edges of the slab are restrained against rotation by hogging strength in the slab ($M_p$) of equal strength to the sagging strength ($M_F$), then the collapse pressure of an all fixed edge slab is *double* that of the simply supported equivalent.

In terms of the parameters $\gamma$, $\alpha$, $\beta$, this means that $\gamma$ will be doubled and so too will $\beta$. Hence for fixed edge slabs, $\beta \geqslant 3.0$. For practical purposes an

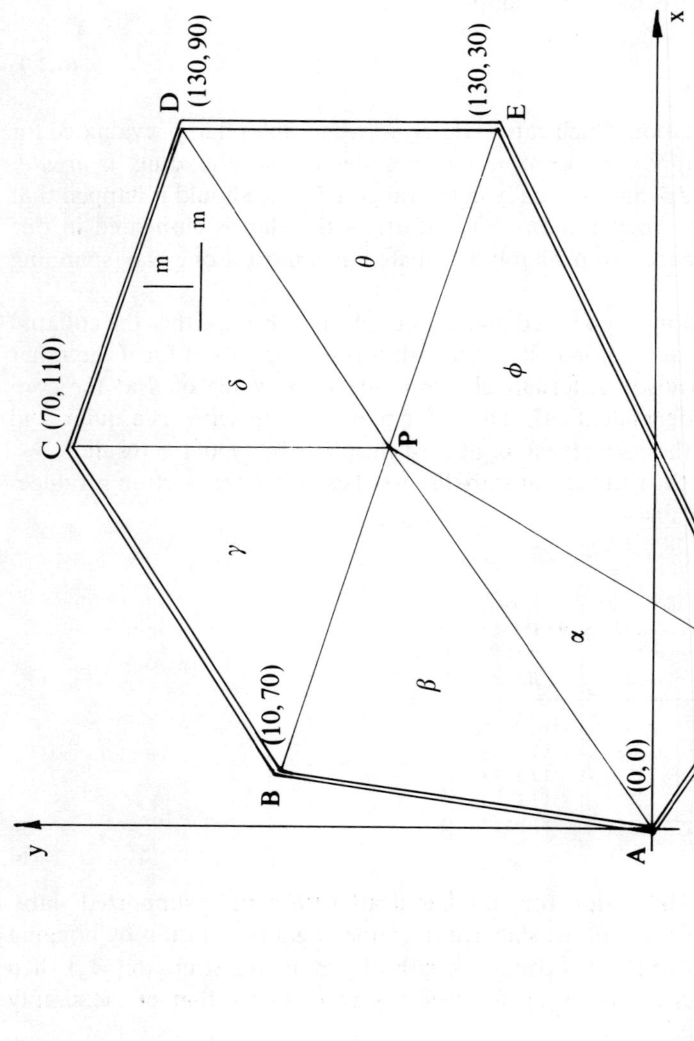


**Figure 4.15** Irregular slab, simple mechanism.

upper limit on $\beta$ is say 3.2, beyond which the slab is sufficiently elongated to be treated as basically a one-way spanning system.

## 4.8 Yield line theory—II. Further isotropic examples

The slab will still be assumed to be isotropic in this section, but a sufficiently complex problem will be investigated to illustrate some of the computational arrangements which might be found useful in the tackling of a typical problem.

The simply supported slab shown in Fig. 4.15 is considered isotropic and loaded with a pressure loading. Assuming the slab strength to sagging moments to be $M_p$ and hence this same value to apply along all yield lines by virtue of the isotropy, it is required to compute a sufficiently close upper bound to be confident that the exact collapse $p_c$ is not being overestimated by more than a few per cent. The intention is to use a work calculation based on an assumed mechanism to calculate an upper bound collapse pressure $p_1$.

Consider first then what a suitable mechanism might be in this case. The slab is very roughly oval in shape; certainly it has a convex shape of edge line, seen from the outside. One possible mechanism is a pyramid shape based on some point (P) in the interior as the apex, with a yield line radiating to each corner. A useful piece of advice in such cases is to begin with the *simplest possible* mechanism in order to generate an answer of some kind. This is much more prudent and satisfying than approaching a more complicated mechanism and possibly even attempting a minimization, becoming bogged down in symbols and formulae and finding the enterprise impossible to complete. So let us begin with this simplest, though possibly rather crude approximation to the problem.

Where then to place the apex? For simplicity let this point (P) be on the join of vertices B and E. Let it be the point with coordinates (70, 50). The analysis of the problem will be written in terms of the actual dimensions. It is advised that the reader redraw Fig. 4.15 on graph paper and follow the steps by making check measurements at the various stages of the calculation.

The six segmental areas into which the slab is divided by the yield lines have been labelled $\alpha$–$\phi$. Measurements of the lengths of the supported edges give the values 36, 70.5, 72, 63, 60 and 112, for $\alpha$–$\phi$ respectively. Next the perpendicular distances of the apex (70, 50) from the respective edges are measured to be 80, 62, 50, 56.5, 60, 44. With these data the plan area of the slab can be evaluated as

$$\tfrac{1}{2}[\Sigma\, 36 \times 80 + \ldots] = 11{,}300 \equiv A.$$

The boundary perimeter, $B = \Sigma l_i = 413.5$.

Then the load work done $= \frac{1}{3}p\Delta A$
and the moment work absorbed $= M_p\Sigma(l_i\theta_1)$

$$= M_p\Delta\left[\frac{36}{80}+\frac{70.5}{62}+\frac{72}{50}+\frac{63}{56.5}+\frac{60}{60}+\frac{112}{44}\right]$$

$$= 7.69 M_p\Delta.$$

$$\therefore \frac{pA}{M_p} = 3\times 7.69 = \underline{23.1}.\ (=\gamma).$$

Now

$$\alpha \equiv \frac{B^2}{A} = 15.11 \quad \therefore \beta = \frac{\gamma}{\alpha} = \frac{23.1}{15.1} = 1.53.$$

Note the closeness of $\beta$ to the lower limit of 1.50. This suggests that the point (70, 50) as apex is nearly the most advantageous choice. As a check, another vertex can be chosen, and the calculation quickly repeated. For example, if the point (50, 40) is chosen then the corresponding estimate for p is given by

$$\gamma = \frac{pA}{M_p} = 3\times 7.89 = \underline{23.7},$$

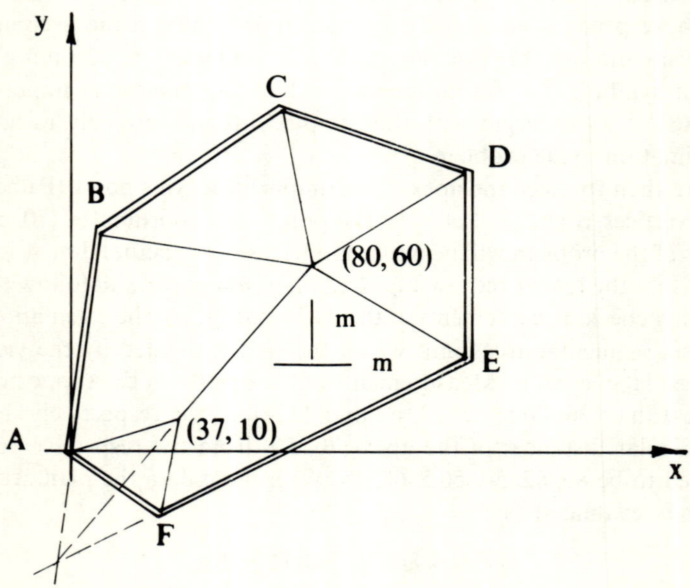


**Figure 4.16** Example—second choice of mechanism.

and hence

$$\beta = \frac{\gamma}{\alpha} = \frac{23.7}{15.1} = 1.57.$$

This again confirms that the initial choice of point P (70, 50) was a sound one.

To complete the investigation of this example some other plausible mechanism should be chosen. This is left as an exercise.

*Exercise:* Repeat the above example but with the choice of mechanism as shown in Fig. 4.16. This is still a one degree of freedom mechanism, but the calculation is a little more complicated than the previous mechanisms associated with this shape. *Answer:* $\gamma = 23.4$; $\alpha = 15.5$; $\beta = 1.55$.

## 4.9 Yield line theory—III. Orthotropic problems

Thus far only isotropically reinforced slabs have been considered. If the reinforcement provided is *different* in the two orthogonal directions, then the system is no longer isotropic. Such slabs are referred to as *orthotropic slabs*. The yield line calculations are similar to those for isotropic slabs, but it must be remembered that now the moment strength along any particular yield line will depend upon how this line is oriented with respect to the now different slab strengths in the two orthogonal directions. Figure 4.11 (p. 97) shows a typical yield line crossing orthotropic reinforcement. The relevant yield moment about this line is given from (4.7.0) as

$$m_\theta = m_2 . \mathrm{Cos}^2\,\theta + m_1\,\mathrm{Sin}^2\,\theta. \tag{4.9.0}$$

If $m_2 = \mu m_1$ then $\mu$ will be termed the *orthotropic ratio*. Generally $\mu$ will be in the range $0.25 \leqslant \mu \leqslant 4$ with the isotropic ratio providing a useful comparison value at $\mu = 1$.

Consider again the example (4.7) of the simply supported rectangular plate, loaded with a uniform pressure, but now reinforced orthotropically. Figure 4.17*a* is a redrawn version of Fig. 4.13*b*, with L = 2l and $m_1 \neq m_2$. Calculation of the collapse pressure can proceed exactly as for the isotropic case, except that now the yield moment along the yield lines such as EC must be calculated from (4.9.0). In the present case, suppose $m_2 = m_1/2$, i.e. $\mu = 0.5$, then the yield moment along a typical line such as EC is given by

$$m_\lambda = m_1\,\mathrm{Sin}^2\,\lambda + m_2\,\mathrm{Cos}^2\,\lambda = m_1(\mathrm{Sin}^2\,\lambda + 0.5\,\mathrm{Cos}^2\,\lambda). \tag{4.9.1}$$

Recall that $\lambda$ is an unknown, to be found from the condition that the collapse pressure (p) associated with a particular $\lambda$ should be a *minimum*. While it is in principle possible to proceed with an algebraic investigation of such problems, it is strongly recommended that instead, likely values of $\lambda$ be chosen, the corresponding p values calculated and the *least* p then chosen as

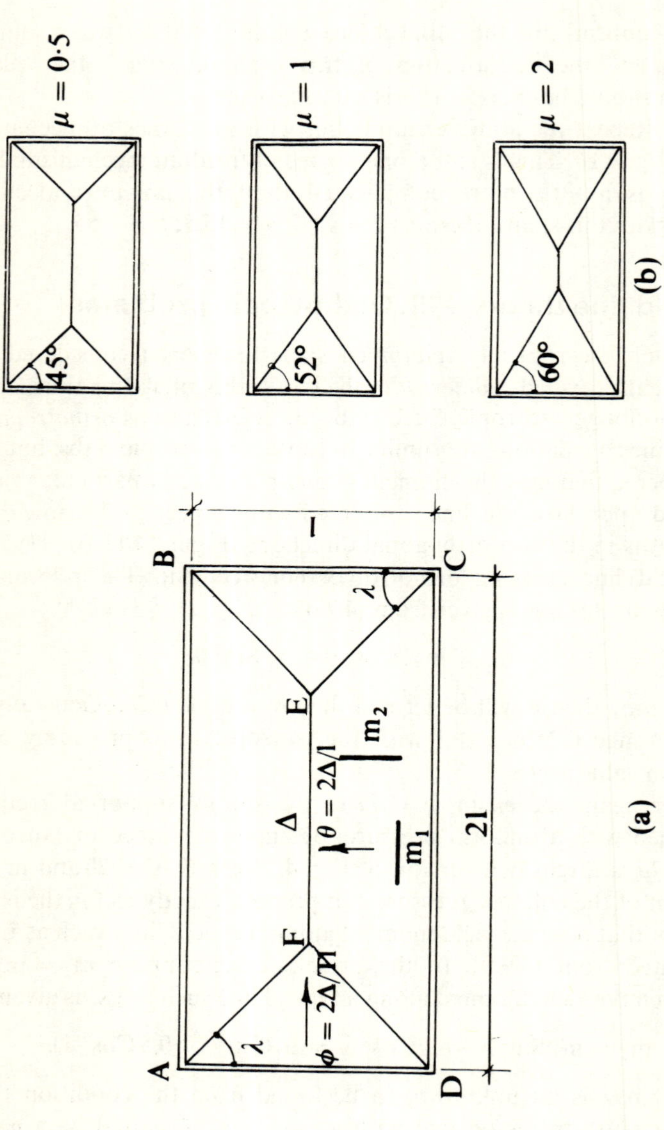


**Figure 4.17** Orthotropic rectangular slab.

the relevant value. Such a method is less elegant but much more likely to systematically identify the minimum value.

It is useful to keep in mind an *associated isotropic* slab—namely that slab which contains the same amount of reinforcement as the orthotropic one. To a good approximation the *amount* of reinforcement is *proportional* to m, the moment yield strength. Also recall that m is expressed per unit width of slab.

Hence the total amount of reinforcement for an isotropic slab is proportional to

$$2M_p \times \text{slab area} \equiv V_M. \tag{4.9.2}$$

This result follows because there are *two* directions of reinforcement and each direction requires an amount proportional to the slab area.

For the orthotropic slab

$$V_M = (m_1 + m_2) \times \text{slab area}. \tag{4.9.3}$$

The symbol V here, and in Chapter 5, is used to denote the total amount of reinforcement.

Hence

$$m_1 + m_2 = 2m \tag{4.9.4}$$

is the required equivalence relation. Now $m_2 = \mu m_1$.

Thus

$$m_1 = \frac{2m}{1+\mu},$$

$$m_2 = \frac{2\mu m}{1+\mu}, \tag{4.9.5}$$

for an orthotropic slab to contain the same amount of reinforcement as the equivalent isotropic slab. In the present case $\mu = 0.5$ and hence

$$m_1 = 1.33m, \quad m_2 = 0.67m. \tag{4.9.6}$$

For the mechanism shown in Fig. 4.17(*a*)

$$4m_1(2-T) + 4m_\lambda . (T + 1/T) = pl^2(1 - T/6), \tag{4.9.7}$$

where $T = \text{Tan}\,\lambda$. Then (4.9.7) is the orthotropic equivalent of (4.7.3).

What is recommended is that trial values of $\lambda$ are used to evaluate p. For example, if $\lambda = 45°$ then from (4.9.1), (4.9.6) $m_1 = 1.33m$, $m_\lambda = 0.75 \times 1.33m = m$. But $T = 1$ and hence from (4.9.7) $pl^2/m = 16$. Other values of $\lambda$ can be chosen and the evaluation of p repeated. Thus $\lambda = 40°$ leads to $m_\lambda = 0.94m$ and $pl^2/m = 16.1$, while $\lambda = 50°$ gives $m_\lambda = 1.05m$ and $pl^2/m = 16.1$, also. There is evidently a very shallow minimum for p at about $\lambda = 45°$.

What then has been established? It has been shown that if the reinforcement in the rectangular slab, side ratio 2 : 1, is arranged with the moment strength *across* the shorter span to be *double* that along the slab length, namely $\mu = 0.5$, but with the total amount of reinforcement constrained to be the *same* as that required for an isotropic slab of strength m, then

$$\left(\frac{pl^2}{m}\right)_{\mu=0.5} = 16, \left(m_1 = \frac{4}{3}m, m_2 = \frac{m_1}{2}\right)$$

and from Table 4.0,

$$\left(\frac{pl^2}{m}\right)_{\mu=1} = 14.1. \tag{4.9.8}$$

In other words, the orthotropic slab can sustain some 13% more load than the associated isotropic slab. There is a catch here, however. Suppose instead of $\mu = 0.5$ the ratio $\mu = 2$ had been used, then the same amount of reinforcement is implied but now

$$m_2 = 1.33m, \quad \text{and} \quad m_1 = 0.67m.$$

The roles of the two directions are exchanged and

$$m_\lambda = m_2(\text{Cos}^2\lambda + 0.5\,\text{Sin}^2\lambda). \tag{4.9.9}$$

For $\lambda = 45°$ the value of $m_\lambda$ is exactly the same for $\mu = 0.5$, namely $m_\lambda = m$ but the associated collapse pressure from (4.9.7) is less, $pl^2/m = 12.8$. The minimum value of p is now when $\lambda = 60°$ with $pl^2/m = 11.85$.

Hence the comparison to be made is

$$\begin{aligned} \mu &= 0.5, \quad pl^2/m = 16, \\ \mu &= 1, \quad pl^2/m = 14.1, \\ \mu &= 2, \quad pl^2/m = 11.85. \end{aligned} \tag{4.9.10}$$

The $\mu = \frac{1}{2}, 2$ cases are coupled since they imply equal total amounts of reinforcement, and this amount has been chosen to be the same as for $\mu = 1$, the associated isotropic case. Hence it is seen that there is a correct and an incorrect way to introduce orthotropy. In the present case it is easy to conclude on intuitive grounds that $\mu = \frac{1}{2}$ is to be preferred to $\mu = 2$, but in more complicated examples this may not be so easy. It is suggested therefore that the associated isotropic value of $\mu = 1$ should always be computed and then used as a measure of the advantage to be gained by the proposed orthotropic layout. Should it happen that the orthotropic strength is *less* than the isotropic one, then change the ratio to be the inverse of the previous value and repeat the calculation. This will lead to a strength advantage for the new orthotropic arrangement.

In Fig. 4.17*b* are shown the positions of the critical yield lines for the three values of $\mu$ studied. As the short span is made stronger at the expense of the long span, so the system becomes more *one* dimensional and stronger, with the corner lines being confined to a smaller area of the slab. If even more extreme values of $\mu$ are chosen it can be shown that a maximum (theoretical) advantage is gained at around $\mu = 0.1$ of about 24% pressure increase for the orthotropic slab. However, in practical cases such a ratio would probably lead to unsightly cracking parallel to the short side in the end regions.

By adopting the system of trial choices for $\lambda$ and evaluation of p, systems of almost any complication can be handled. The reinforcement need not be orthogonal, and different ratios may be adopted for the bottom and top steels. An alternative and more elegant approach would have been to introduce the *affinity* theorems of Johansen (see chapter 6). There are limitations to use of the theorems however and some aspects of the physical situation are lost in the process, and so these theorems have been omitted.

One final comment on orthotropy—the non-dimensional parameters $\alpha$, $\beta$, $\gamma$ introduced in (4.7) can be applied to orthotropic slabs, through the associated isotropic slab. Thus for the $\mu = 0.5$ case

$$\gamma = \mathrm{pA/m} = 2 \times \mathrm{p}l^2/\mathrm{m} = 32,$$

and

$$\alpha = (6l)^2/2l^2 = 18.$$

Hence

$$\beta = \frac{\gamma}{\alpha} = 1.78 > 1.5.$$

On the other hand the $\mu = 2$ example gives

$$\gamma = 23.70, \quad \alpha = 18,$$

and hence

$$\beta = 1.32 < 1.5.$$

That $\beta < 1.5$ is another signal showing that $\mu = 2$ is a disadvantageous orthotropic ratio. Instead $\mu = \frac{1}{2}$ should be chosen, with no reinforcement penalty, but a strength advantage.

## 4.10 Strip method—Hillerborg's proposals

In recent years a further method has been gaining popularity in dealing with slab design questions. The initial proposals were made by A. Hillerborg in about 1960 and have been elaborated since. Here only a brief sketch is possible and all the discussion will be limited to rectangular slabs.

From a practical viewpoint, rectangular slabs are likely to be reinforced along directions parallel to the slab edges, even although these directions are unlikely to be principal moment directions. From what has been seen of slab problems so far, it will be evident that many comparatively simple and practical problems cannot be fully solved whether elastic or plastic. In the case of elastic plates, there are only a few shapes capable of analytical solutions. Most data have been accumulated through numerical solutions, notably finite element analyses in recent years. In the case of plastic plates, upper bounds on the collapse pressure are readily found but few associated lower bounds are known.

The Hillerborg proposals aim to resolve this latter stalemate by *assuming* that the (rectangular) slab does not support *any* twisting moment in the x, y directions. In effect the slab is notionally thought of as composed of a grid-work of beams which in some manner interact with one another to carry the imposed loading.

What is aimed at is an equilibrium system of bending moments which support the given load, and hence the method is in the spirit of a lower bound on the collapse pressure. In the simplest form the method assumes all the load is dispersed to the boundaries by beam strips in both x and y directions (Fig. 4.18). A decision is needed on how the load should be divided up. On Fig. 4.18 are shown some dotted lines and arrows. The dotted lines divide the whole area up into zones, and the *whole load* within each zone is then *assumed* to be carried by strips in the direction of the arrow. Some judgement is required in making this subdivision. For the rectangular slab illustrated, $\theta = \pi/4$ is a

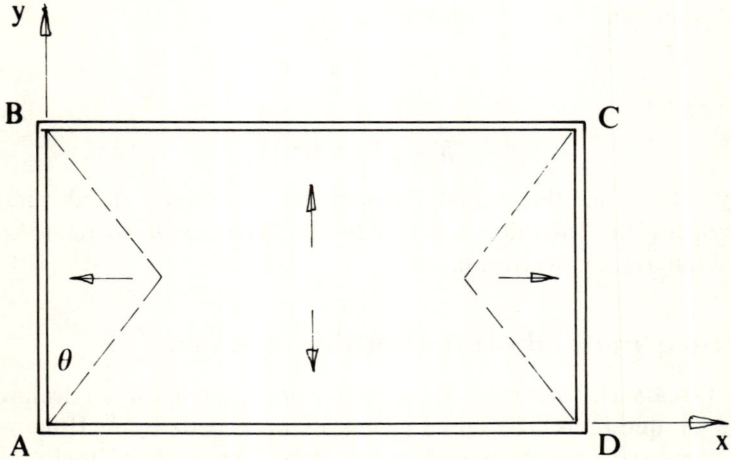


**Figure 4.18** Strip theory—spanning proposals.

reasonable choice. This choice is backed up by prior knowledge of the corresponding elastic solution.

Now a typical x direction beam strip will have loading near its ends with none in the central section, whereas a typical y direction beam strip will carry a full load if near the slab centre, or end loading only for a strip near the ends, at AB or CD.

The dotted lines are sometimes referred to as *lines of discontinuity*. With the discontinuity lines shown in Fig. 4.18, no two adjacent beam strips will have similar loading, and this implies continuously variable reinforcement patterns.

A more practical choice of discontinuity lines is to step them so as to give finite width beam strips which have similar loading. Yet another alternative available is to suppose the total loading (p) to be split locally into two components, $\alpha p$ and $p(1-\alpha)$, which are applied to the beam strips in say x and y directions respectively. The simplest such scheme is to take $\alpha$ to be 1 or 0 as is implicit in Fig. 4.18. A more elaborate variant is to think of $\alpha$ as $\alpha(x, y)$ which can be adjusted across the slab for nearness to a boundary, for example.

In actual use, a balance must be struck between simplicity and useful elaboration. Also, for practical purposes there are limits beyond which the strip solution is likely to produce unacceptable cracking in a real construction. For example, in Fig. 4.18 if $\theta$ is assumed too small then the y spanning strips will be provided near to the short edges and cracking along these edges will occur in a real construction.

Hence unlike the other methods described, the strip method has an important element of discretion, or choice, associated with it. To the inexperienced this signals caution. To the more experienced this method provides a malleable method which can be made to cope with the unusual situation.

The content of the next chapter can be thought of as a systematic development of this theme of strip methods.

## 4.11 Conclusions

In a number of respects, the contents of this chapter, more than any other, are central to current plate problems. Certainly, more progress can be made with many of the problems posed here than is the case with the majority of elastic problems. Some of the more general statements concerned with the non-dimensional ratios offer worthwhile distillations and begin to provide a useful categorization of solutions.

In the next chapter some more modern and radical theory is outlined. This theory may be a growing point in the subject in future, though only time will tell.

# 5 Optimal plates

## 5.0 Introduction

The emphasis in the previous chapters has been on the study of isotropic plates. As very many practical plates are isotropic, the emphasis on isotropy is reasonable, but there are also valid reasons for making a separate study of anisotropy, which is the central theme of this chapter.

From the practical point of view, all the plates studied in this chapter should be thought of as reinforced concrete plates (slabs), since the anisotropy is physically most easily introduced by varying the amount of reinforcement in chosen directions in the plate.

An independent reason why attention should be paid to the methods discussed in this chapter is that they can be used to solve, in closed form, a technically interesting range of anisotropic plate problems. It will already be evident that many interesting and apparently simple isotropic plate problems cannot be comprehensively solved by the classical methods of chapters 3 and 4. Hence any (non-classical) method of formulation and solution which allows many closed form solutions to be obtained merits some study. This is the situation with *optimum plate theory* studied in this chapter.

## 5.1 Problem formulation

The basic aim is to investigate layouts of reinforcement for reinforced concrete, or other fibre-reinforced composites, such that the amount of reinforcement needed to support the load is a *minimum*, or near minimum. The plates will be assumed to be of constant thickness and loaded with static loads. The bulk of the plate will be composed of the matrix (concrete or other filler), reinforced in the *tension* face of the plate with steel reinforcement (or other fibre such as glass or carbon).

The purpose of the study is to suggest amounts of reinforcement, and where and in what directions these should be placed, to achieve minimum fibre consumption for given load. Alternatively, the theory and methods of solution can be seen to produce maximum load capacity for given fibre content when this is arranged optimally.

Consider first a uniform depth and width concrete beam carrying a uniformly distributed load (p) and simply supported at the ends. The bending moment distribution in this beam is determinate—in this case it is a parabolic variation from zero at the ends to a maximum central (sagging) value of $pL^2/8$, where L is the span. Suppose now that only *just* sufficient (tensile) reinforcement is provided in the concrete to support this bending moment. Then with the assumption that the depth of the concrete stress block is *small* compared with the beam depth, the conclusion to be drawn is that the steel force (=stress ($\sigma$) $\times$ sectional area (A)) must be *proportional* to the bending moment (M). Indeed

$$M = \sigma A d$$

where d is the effective beam lever arm, *assumed constant.* Suppose further that the steel working stress can be held constant at some suitable fraction of the yield stress, then it is concluded that

$$A \propto M.$$

In other words, the cross-sectional area of the (steel) tension reinforcement (A) required is proportional to the local bending moment (M).

Hence the total volume of (steel) reinforcement required to support this bending moment distribution is given by

$$V_R \equiv \int_0^L A\,dx,$$

and

$$V_M \equiv \int_0^L |M|\,dx,$$
$$= V_R \cdot \sigma \cdot d,$$

where $V_R$ is the volume of reinforcement, and $V_M$ is defined to be the *moment volume* associated. Minimizing the moment volume achieves the same result as minimizing the volume of reinforcement, $V_R$. In what follows, most attention will be focused on $V_M$.

In the present example, with the x coordinate measured from one end of the beam,

$$M = \int_0^L |M(x)|\,dx = \int_0^L \left| -\frac{px(L-x)}{2} \right| dx$$
$$= \frac{p}{2}\left[\frac{L^3}{2} - \frac{L^3}{3}\right] = \frac{pL^3}{12}.$$

In order to achieve the condition that $\sigma$, the fibre stress, be a constant, it will be evident that the beam *curvature* must be constant, since $\sigma$ constant implies a constant *strain* in the reinforcement and this can only be achieved from a constant curvature.

This observation about the *constancy of the beam curvature* is the all-pervading flavour of the treatment of optimum plates which follows in this chapter.

Though full proofs of the necessary and sufficient conditions for achieving a minimum of $V_M$, and hence $V_R$, will not be developed, it can be appreciated after a few examples that optimum (or minimum) configuration of curvatures and hence deflected shape, and associated bending moments conform with the following requirements.

First, curvatures and bending moments must *correspond.* By this is meant that sagging curvature must be associated with sagging bending moment. Secondly, the deflected shape must be *smooth.* This means that adjacent constant curvature segments must meet along a common curve and the slope *transverse* to this common curve must be common to the two segments.

Those directions which are reinforced must be directions along which the curvature is *constant.* For if this was not the case, then the reinforcement would not be strained to the constant value desired.

Again the stressed reinforcement determines the internal bending moments and these bending moments must be in equilibrium with the loading. So equilibrium must be achieved, with constant curvatures along the reinforcement directions, and the deflected surface segments must fit together to produce a *smooth* whole. These appear to be very difficult conditions to satisfy simultaneously, with little constructive guidance to assist. But there is considerable scope for intuition to play a part here, as will be demonstrated shortly.

A key element in the search is to *expect* that, very often, the appropriate bending moment system is likely to be *one-dimensional.* The examples will illustrate this observation.

## 5.2 Constant curvature surfaces and principal directions

All the constant curvature surfaces considered here are characterized by constant curvature trajectories which are *straight* when seen in plan. Recall that the deflected surface of a deformed plate is a shallow shape. This implies that the vertical section taken along a *constant curvature direction* will reveal a *shallow circular arc* of section. When described by a local cartesian coordinate system, such a piece of arc will be parabolic since slopes are small and are to be neglected compared to unity.

If the constant curvature has the value k, then point A, a distance d from a zero slope point (B) of a constant curvature arc, will have a transverse displacement

$$w_A = w_B \pm \frac{kd^2}{2}$$

the sign depending upon the sign convention adopted for curvatures and displacements.

Likewise the slopes at A and B will differ by an amount $|kd|$. Both these results are very useful in exploring the constant curvature shapes which will shortly be considered, and are based upon the identity of a *shallow* circular arc with a parabolic curve.

In the Appendix a good deal of attention is paid to the analysis of surface geometry to allow such features as principal directions on surfaces to be analysed. Although principal directions play a vital role in optimum plate studies, the analysis needed to facilitate this study is quite minimal. The essential feature of a principal direction (refer to Section A.3, p. 152) is the property that if two adjacent points on the surface are considered, and the normals to the surface are drawn at these two points, then the *direction* defined by the *join* of these points is a *principal direction* if and only if these two adjacent normals are *coplanar*. Here, consideration is being restricted to *straight* constant curvature directions, and it is these directions which are intended to be principal directions.

A very important theorem will now be stated which identifies when these straight, constant curvature directions will be principal directions on the shallow deflected slab surface. Consider two points (1, 2), not necessarily adjacent, on a straight constant curvature direction. Generally at each of these points the surface in the transverse direction will have some slope, which will be denoted by $S_1$, $S_2$ at the two points 1, 2 respectively. The theorem states that if a given straight trajectory on a shallow surface is a *constant curvature principal* direction, then the slope transverse to this direction is the *same* at *all* points. This result follows from the requirement that adjacent normals be coplanar. Consider the use of the theorem in the quest for *principal constant curvature* directions. To be principal, adjacent normals must be coplanar. This requires that the S values at the adjacent points must be the *same*, since otherwise, viewed with the joining piece of arc end-on, the normals would be skew. Now apply the theorem and note that if the S value is the same at *two adjacent* points it is the same at *all* points. Hence a necessary and sufficient condition for a straight shallow constant curvature arc to be a *principal* direction on the surface on which it is scribed is that the value of the transverse slope S at all points be the same. This result is of fundamental importance in determining whether surfaces and directions are suitable for satisfying the optimal requirements.

To summarize then, the constant curvature surfaces of interest in the theory of optimal plates are shallow surfaces possessing straight constant curvature directions which are principal directions on the deflected slab surface. Such directions are characterized by the property that along any particular trajectory the surface slope *transverse* (that is at right angles) to the constant curvature direction is *constant*.

## 5.3 Basic results—corners

This introduction to optimal plate theory will be presented from an intuitive viewpoint to begin with; then some of the supporting theory will be discussed. Here it is proposed to examine the layout of optimal, constant curvature directions in the neighbourhood of intersecting edges. From these beginnings, whole problems can later be solved.

The guiding consideration in the intuitive, physically motivated study of these problems is that the optimal spanning layout is likely to be *one-dimensional*. In other words, locally, any optimal zone will resemble a series of constant curvature beam elements. This tendency for such layouts to be one-dimensional has been suggested by theory and will be discussed later in a more formal manner.

Once this tendency has been identified, several solutions for corner zones quickly follow. Consider first a piece of slab between two intersecting simply supported edges.

### (a) *The simply supported corner*

The aim is to describe both a possible spanning arrangement and a deflected shape for the slab in constant curvature terms, in the neighbourhood of the simply supported boundaries.

Consider the strip AB drawn normal to the angle bisector. This strip can support any distribution of superimposed load and self weight, by suitable simple beam reinforcement. Also if this beam is bent to a constant sagging curvature of $-k$, then all this reinforcement will be strained to a constant amount and hence to a constant stress level. These features are the ingredients of an optimal system.

While it has not yet been proved it will be evident from considering alternative arrangements that the series of parallel spans, of which AB is a typical one, provide the optimal layout. Taken together the series of parallel spans generate a rather simple constant curvature surface. From symmetry it is true that AB is a principal direction on this surface. Also the second principal curvature, namely that curvature which is normal to the constant curvature along AB can be evaluated as follows.

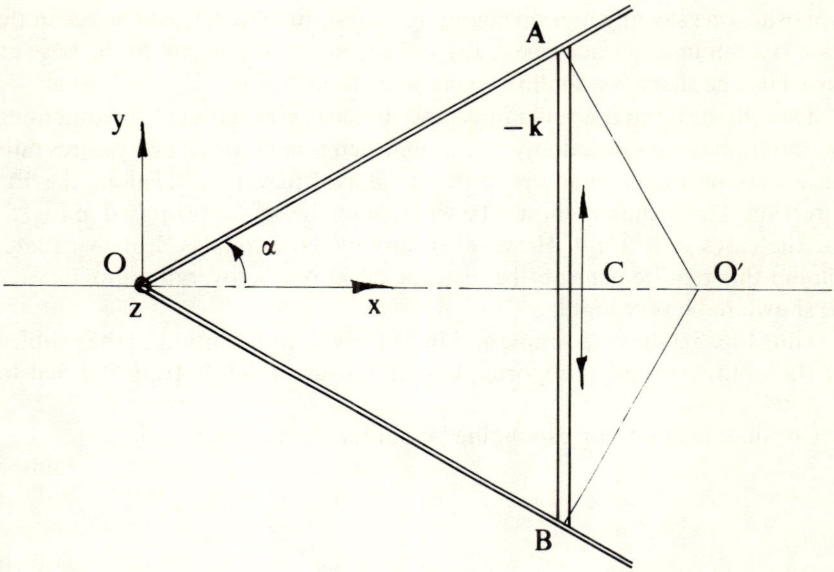


**Figure 5.0** Simply supported corner arrangement.

Set up a cartesian system of coordinates at the corner (Fig. 5.0). Then with an included angle of $2\alpha$,

$$AC = x \,.\, \operatorname{Tan} \alpha.$$

Hence, using the expression from 5.2 for a constant curvature arc, since C is a zero slope point, the difference in transverse displacement between A and C is given by

$$w_c = \frac{k}{2}(AC)^2,$$

since $w_A = 0$ from being on the simply supported edge.

Thus

$$w_c = \frac{k}{2} x^2 \operatorname{Tan}^2 \alpha.$$

The second principal curvature, $\kappa_2$, is then given by

$$\kappa_2 = \frac{d^2w}{dx^2} = k \operatorname{Tan}^2 \alpha. \tag{5.3.0}$$

From this expression it is seen that $|\kappa_2| \leqslant k$ so long as $\alpha \leqslant \pi/4$. When $\alpha = \pi/4$ then $|\kappa_2|$ is equal to k and the principal curvatures are equal and opposite, one sagging and one hogging. The optimal deflected surface in this case is a minimal surface (see A.13) and sections taken at $\pi/4$ to the edge are *straight*. The shape is a shallow hyperbolic paraboloid.

Though the point is not of importance in the present context, a requirement of optimum theory is that any direction which is *not* reinforced by fibres must have a curvature *less* than the constant value of curvature $|k|$ along the fibre direction. This requirement will be violated by the layout proposed in Fig. 5.0 for the cases of $\alpha > \pi/4$. However in any of the examples dealt with later, should this requirement not be met, the effect on the overall optimality can be shown to be very slight.

Thus Fig. 5.0 shows a simple and intuitively sound solution to the problem of the optimal simply supported corner layout, which is truly optimal for $\alpha \leqslant \pi/4$.

Consider next the corresponding layout for the clamped corner.

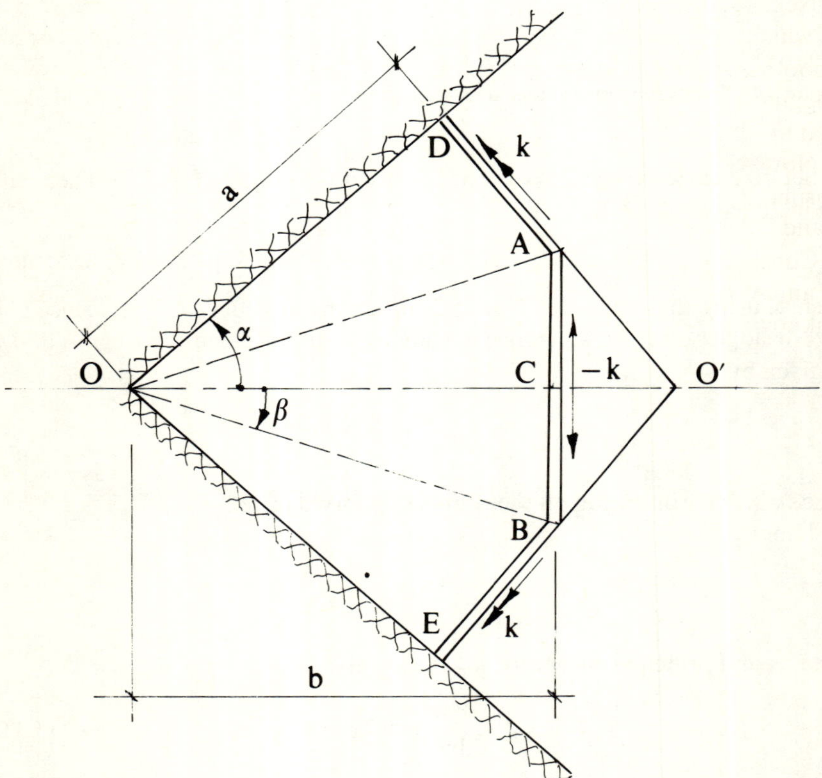


**Figure 5.1** Clamped corner arrangement.

(b) *The clamped edge corner*

In this case the only readily obvious feature of the surface shape and spanning layout is that it should have symmetry about the angle bisector.

The clamped edge suggests cantilever spans in the neighbourhood of the edges, with possibly an infilling simply supported span. But what are the proportions and details? These must be found from the requirement that the deflected surface so generated by constant curvature hogging cantilever and sagging simply supported spans, *must* describe a *smooth* deflected surface.

Sketched in Fig. 5.1 is a tentative layout of cantilever beam strips spanning out normally from the edges, and joined by the simply supported span AB which is thought of as similar to the span AB of the simply supported corner. Then DA and EB will have hogging curvature k down into the figure and AB will be a sagging span of constant curvature $-k$ and zero slope point at mid-point C, as demanded by the overall symmetry with respect to the angle bisector.

Next, steps must be taken to ensure that the whole series of adjacent spans, of which DACBE is a typical example, when fitted side by side, do form a smooth surface. The essential condition to examine is the intersection of hogging and sagging spans at the typical point, A. The procedure will be to examine the surface slope vectors for the hogging and sagging sections.

Now DA is a segment of a circular cylinder with zero slope along OD and normal to OD. Hence the slope vector at A for DA is given by the product of k and DA (Fig. 5.2) and is directed along DA.

Consider now CA. The point C is a zero slope point and ACB is a segment of an as yet not fully defined surface. Across ACB there will be some as yet unknown surface slope S, which will be constant if ACB is to be a principal

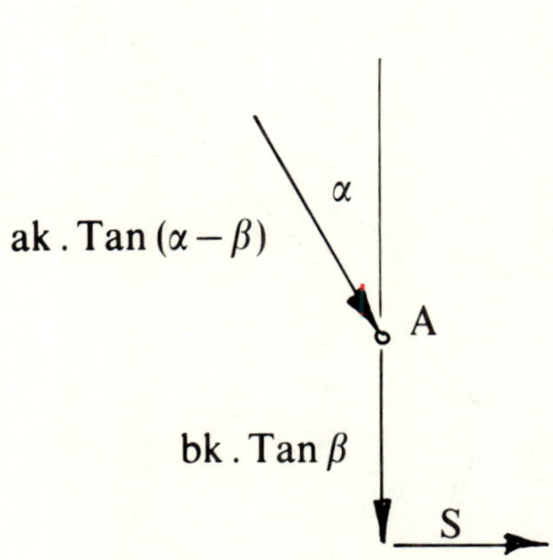


**Figure 5.2** Slope conditions at a junction.

direction. The other component of surface slope at A for ACB will be $k \times AC$, since C is a zero slope point, when seen in elevation on ACB.

All this slope information is shown in Fig. 5.2. As surface slope is a *vector* quantity, the vector for portion DA must be identical to the vector for portion ACB, in order that there be surface slope continuity at A. The requirements are that any two selected non-parallel components must be common. Hence

$$bk \operatorname{Tan} \beta = ak \operatorname{Tan}(\alpha - \beta) . \cos \alpha \quad (5.3.1a)$$

and

$$S = ak \operatorname{Tan}(\alpha - \beta) . \sin \alpha . \quad (5.3.1b)$$

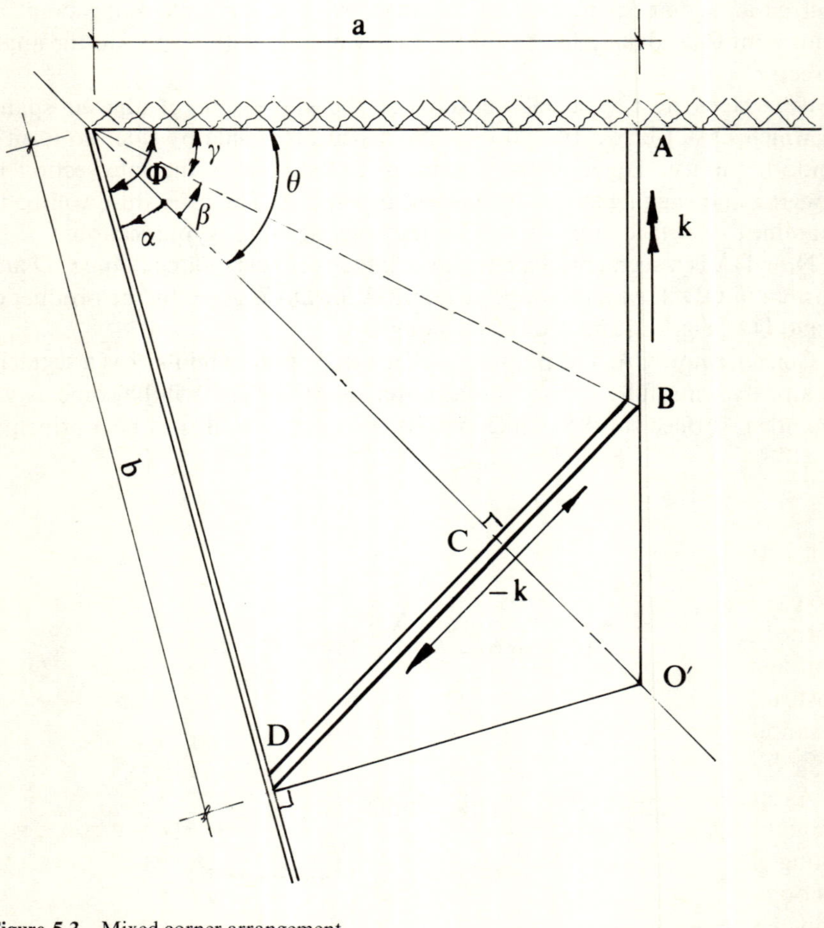


**Figure 5.3** Mixed corner arrangement.

Now for some manipulation and simplification. From the geometry of Fig. 5.1,

$$a\,\mathrm{Sec}\,(\alpha-\beta) = \mathrm{OA} = b\,\mathrm{Sec}\,\beta$$

or

$$a\,\mathrm{Cos}\,\beta = b\,\mathrm{Cos}\,(\alpha-\beta).$$

Thus (5.3.1a) gives

$$b\,\mathrm{Sin}\,\beta = b\,\mathrm{Cos}\,(\alpha-\beta)\,.\,\mathrm{Tan}\,(\alpha-\beta)\,.\,\mathrm{Cos}\,\alpha = b\,\mathrm{Sin}\,(\alpha-\beta)\,\mathrm{Cos}\,\alpha$$

or

$$\mathrm{Sin}\,\beta = \frac{\mathrm{Sin}\,2\alpha}{2}\,.\,\mathrm{Cos}\,\beta - \mathrm{Cos}^2\,\alpha\,.\,\mathrm{Sin}\,\beta.$$

Hence

$$\mathrm{Tan}\,\beta = \frac{\mathrm{Sin}\,2\alpha}{2(1+\mathrm{Cos}^2\,\alpha)}. \tag{5.3.2}$$

The expression for $\beta$, (5.3.2), therefore fixes the geometry of the corner. For example, if $\alpha = \pi/4$ then $\mathrm{Sin}\,2\alpha = 1$, $\mathrm{Cos}\,\alpha = 1/\sqrt{2}$ and $\mathrm{Tan}\,\beta = 1/3$.

*Exercises:* (1) Show that $\kappa_2$, the transverse (principal) curvature in the region OAB, is independent of b, is hogging, and given by

$$\kappa_2 = \frac{k\,\mathrm{Sin}^2\,\alpha}{1+\mathrm{Cos}^2\,\alpha}.$$

(2) Show that the surface transverse slope S in both the simply supported corner and the clamped corner is given by

$$S = k(O'C).$$

This is a very important geometrical result.

### (c) *The mixed corner*

To complete the analysis of elementary corner results, the case of the mixed corner consisting of a simple span meeting a clamped span will now be considered. In this case there is not even symmetry to take as a starting point. Instead, suppose the spanning system is a clamped-edge strip AB, supporting a simple span BC, which in turn is supported on the simply supported edge (Fig. 5.3). This is in the spirit of the previous investigations.

As an extension of these earlier examples the present example can be thought of as a composite of suitable halves from (a) and (b) brought together along their respective lines of symmetry. For these portions to fit together smoothly the transverse curvatures and displacements of the pieces must coincide.

In Fig. 5.3 has been sketched a tentative line of separation, OC, which is half a simply supported corner, semi-angle $\alpha$ combined with a half clamped corner, semi-angle $\theta$. For this to be a smooth junction the angles $\alpha$ and $\theta$ must be in the correct relation to give a common transverse curvature.

From the relations developed above this requires that

$$\mathrm{k}\,\mathrm{Tan}^2\alpha = \frac{\mathrm{k}\,\mathrm{Sin}^2\theta}{1+\mathrm{Cos}^2\theta}.$$

Simplifying,

$$\sqrt{2}\,.\,\mathrm{Sin}\,\alpha = \mathrm{Sin}\,\theta. \tag{5.3.3}$$

Expressing this in $\alpha$, $\Phi$ terms

$$\sqrt{2}\,.\,\mathrm{Sin}\,\alpha = \mathrm{Sin}\,(\Phi-\alpha).$$

Solve for $\alpha$ and obtain

$$\mathrm{Sin}\,\alpha = \frac{\mathrm{Sin}\,\Phi}{(3+2\,.\sqrt{2}\,.\,\mathrm{Cos}\,\Phi)^{1/2}}. \tag{5.3.4}$$

This is the primary relation of this study and allows the whole configuration of Fig. 5.3 to be established.

*Exercises:* (1) Study Fig. 5.3 further, show that O′D is normal to OD, and that

$$\mathrm{Tan}\,\beta = \frac{\mathrm{Sin}\,2\theta}{2(1+\mathrm{Cos}^2\theta)}.$$

A very convenient approach is to draw Fig. 5.3 to scale.

(2) For the case $\Phi = \pi/2$ show that

$$\alpha = \gamma, \quad \mathrm{Tan}\,\alpha = 1/\sqrt{2}, \quad \mathrm{Sin}\,\beta = 1/3, \quad \mathrm{b} = \sqrt{2}\mathrm{a}, \quad \mathrm{AB} = \mathrm{BO}'.$$

Draw the layout to scale.

The other type of edge which might be found in combination with the previous two is the *free edge*. The guiding rule here is that the constant curvature spans will always be *parallel* to the free edge. How this can be achieved will be seen in examples to follow later.

To conclude, in this section we have discussed the likely spanning directions and deflected shapes in terms of associated constant curvature spans for the pieces of slab in the neighbourhood of corners. These arrangements have not yet been shown to be optimal. Indeed, one of the exercises readers can set themselves is to propose possible alternative spanning arrangements, and explore the geometrical consequences and load carrying capabilities of these alternatives.

## 5.4 Some complete results

Here the question of proposing possible layouts for the spanning and deflected shape of *complete* slabs will be initiated. Begin with the simplest case, the simply supported circular plate. Geometrically all the requirements for smoothness of the deflected shape built up from constant curvature spans can be met by a shallow spherical surface with sagging curvature $-k$. This can be interpreted in statical terms to mean that any desired system of beams spanning the region is permissible, for example, a parallel system of beams, or if desired a grid of beams. It will be shown later that the strength required, in terms of reinforcement needed is, in theory, unaffected by the choice. Of more interest is the solution appropriate to a clamped edge circular plate. It is relatively easy to show that the only configuration of hogging edge zone with sagging central zone, constant principal curvatures in the fibre directions and smooth deflected shape, is that composed of an outer hogging zone extending to mid radius and a central spherical zone (Fig. 5.4).

A more practical shape is the rectangle. Consider first the simply supported rectangle in Fig. 5.4. There is a combined deflected surface built up from the corner zone discussed in 5.3, with a spherical deflected surface, labelled N (neutral), and a purely cylindrical portion. All these segments imply reinforcement in the direction of the constant curvature, $|k|$, and it can be seen that the resulting spanning system is in effect a one-way grid of beams, except in the neutral zone where a two-way grid is theoretically acceptable (since all directions are principal). But smoothness of the deflected surface is achieved, even if the rectangle is distorted into a diamond shape.

The next stage is to consider the clamped rectangle in Fig. 5.4. Now the layout looks to be significantly more complex but the same fitting together of corner zone, neutral zone and cylindrical portions leads to the layout shown, and this can be demonstrated to possess all the necessary features.

Next in this group of complete solutions can be demonstrated the application of the mixed corner layout developed in the previous section to incorporation into the rectangular slab with alternate edges simply supported, clamped or free. Once again the elements are the same—the corner, the neutral zone and a cylindrical segment. And once again the geometrical properties ensure the smooth fitting together of the pieces.

As typical of the further application of the basic elements of solution via corner geometry to solve other problems, consider the triangular plate in Fig. 5.4. Now the vital role that O′ plays can be best appreciated. This point is common to the three corners, and is the incentre for the triangle. Once again,

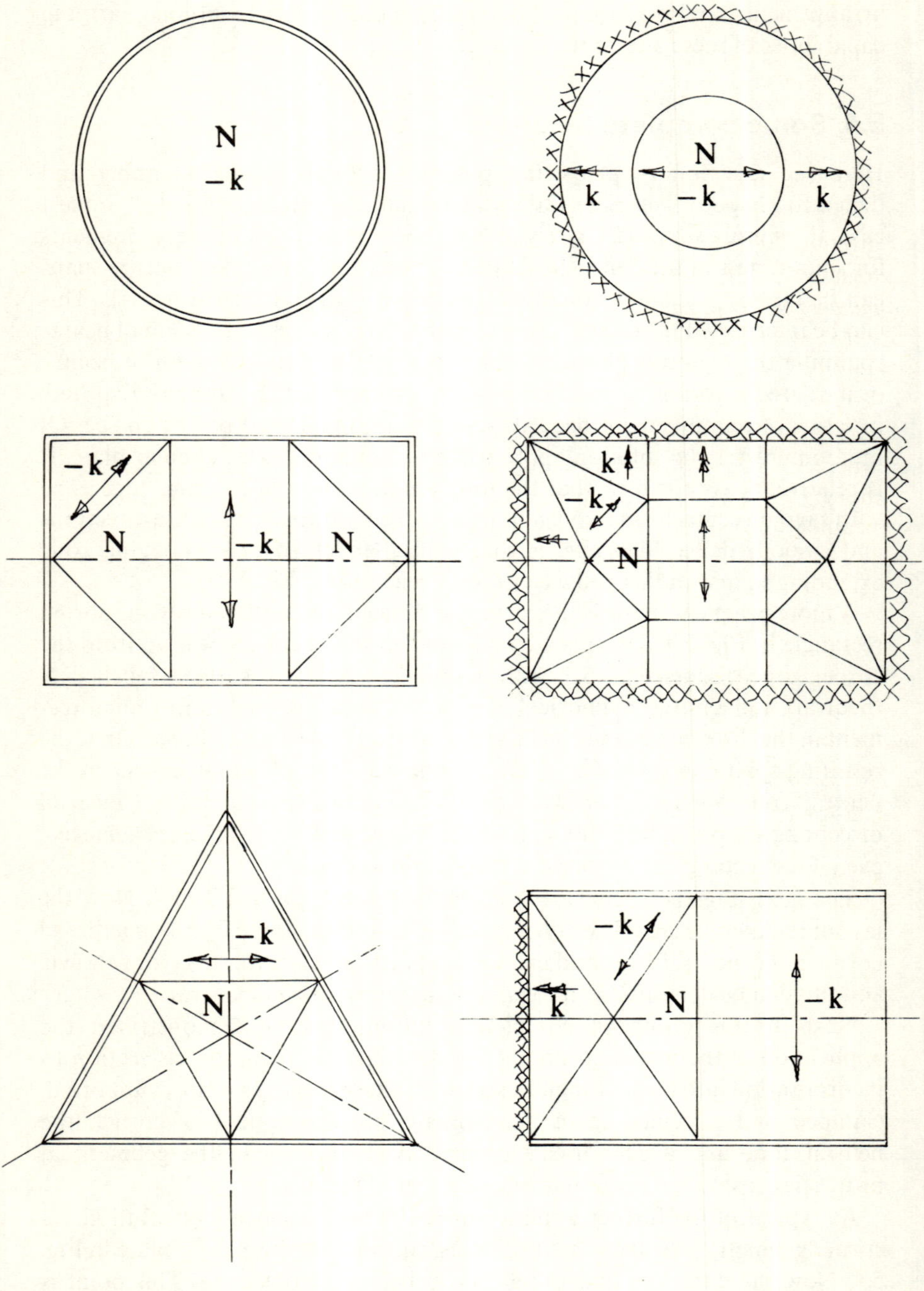


**Figure 5.4** A selection of optimal layouts.

the essentially elementary geometrical properties are the key to constructing a possible spanning and deflection layout.

## 5.5 Moment volumes

The measure of optimality used here is the amount of reinforcement needed to support a given load on a slab spanning to the given boundaries. Because attention is to be confined to constant depth slabs, it is convenient to formulate the measure of optimality as the quantity defined earlier, the *moment volume*, rather than the actual *reinforcement volume.*

Moment volume is defined as

$$V_M \equiv \int (|M_1| + |M_2|)\,dA$$

where $M_{1,2}$ are the principal moments, and dA is the element of slab (plan) area. This expression is analogous to the expression defined earlier for a beam. If needed, the reinforcement volume ($V_R$) can be found from

$$V_R = \frac{V_M}{\sigma d},$$

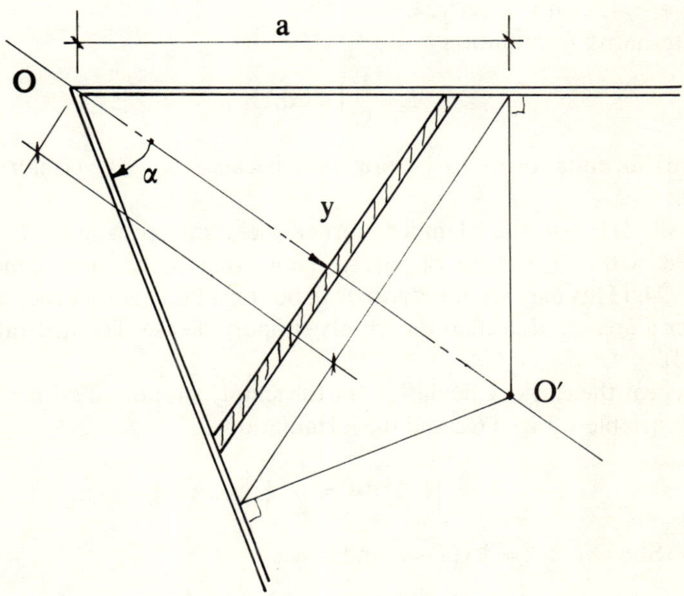


**Figure 5.5** Simply supported corner—moment volume.

where $\sigma$ is the common stress in all reinforcement and d is the effective lever arm for all reinforcement, which is assumed to be constant. When the context is clear, the subscript (M or R) will be omitted for simplicity.

As has been emphasized in the construction of feasible spanning arrangements in 5.3 and 5.4, the most prominent feature of the layouts is the local beam-like quality of the problem. The only portion of the layout where this may not be the case is in the neutral zone. Indeed, within the neutral zone the spanning arrangement can be chosen to suit any possible requirements. Elsewhere the spanning arrangement is fixed, since the reinforcement must lie in the direction of the principal curvature, $|k|$.

Consider the moment volume calculation for the simply supported corner (Fig. 5.5). Now, with a uniform loading of p applied, including self weight, the moment volume is given by

$$
\begin{aligned}
V &\equiv \int |M|\, dA \\
&= \int_0^{y \operatorname{Tan} \alpha} \int_0^{y} \frac{p}{2} (y \operatorname{Tan} \alpha - x)(y \operatorname{Tan} \alpha + x)\, dy\, dx \\
&= \frac{p}{6} \operatorname{Tan}^3 \alpha \, . \, y^4.
\end{aligned}
$$

When $\alpha = \pi/4$, then $V = pa^4/24$.

An alternative formulation is

$$V = \frac{1}{k} \int pw\, dA.$$

Often this formulation is to be preferred because an easy visualization is possible.

*Exercises:* (1) For the clamped corner including an angle of $\pi/2$ and contained within a square of side a, show that V has the same value, $V = pa^4/24$. (This may seem paradoxical, but recall that this corner supports load over a greater area than the simply supported case. The neutral zone is excluded).

(2) Repeat the earlier calculation for the simply supported corner, but use now the variable x, Fig. 5.6*a*, and the formulation

$$V = \int |M|\, dA = \frac{1}{k} . \int pw\, dA,$$

to find V. Show that $\Delta = kx(a-x)$ and hence

$$V = \frac{p}{k} \int_0^a \frac{kx}{2} (a-x)^2\, dx = \frac{pa^4}{24}.$$

(3) Repeat the clamped edge V calculation using the variable x, Fig. 5.6*b*. Then

$$\text{area}_{\text{Section A}} = \frac{kx^2}{12}\cdot(6a-5x),$$

$$\text{area}_{\text{Section B}} = \frac{k}{48}(a^3+6a^2x+12ax^2-28x^3).$$

Use these results to complete an independent calculation of V.

## 5.6 Some theory

Thus far the moment volumes computed have not been demonstrated to be minima—the arguments have been intuitive. The aim now is to explore briefly the theoretical reasons why these values are minima.

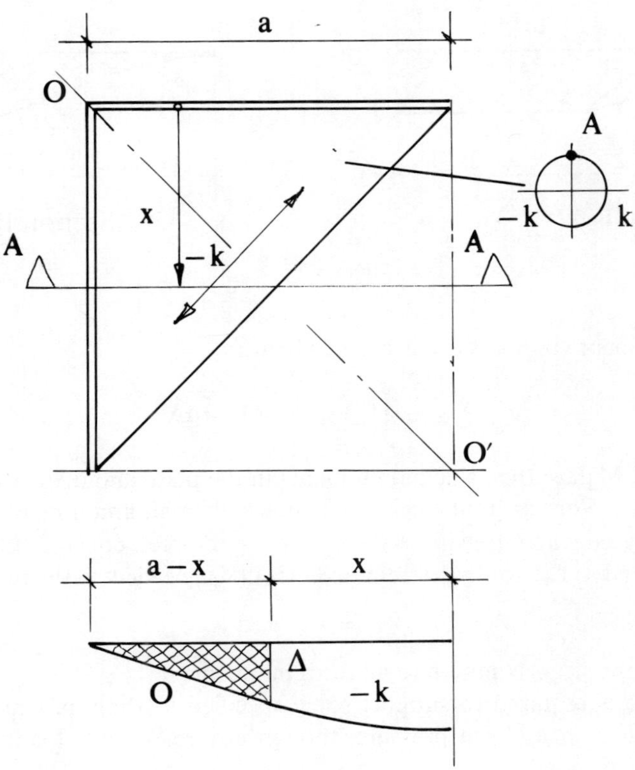


**Figure 5.6**(*a*) Simply supported corner—alternative section.

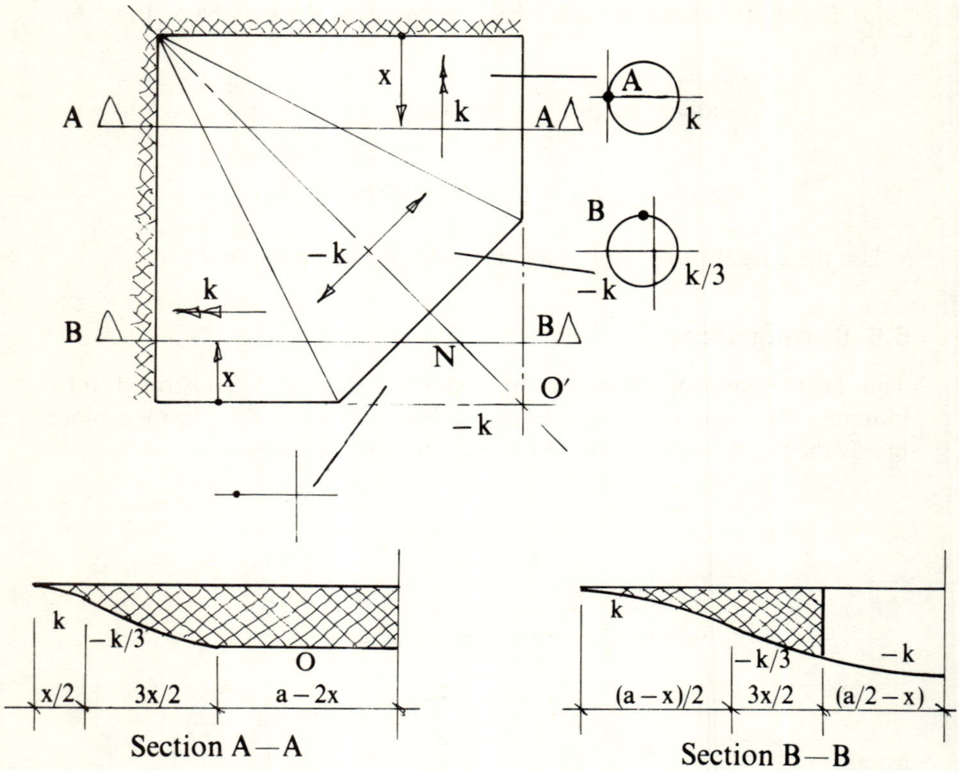


**Figure 5.6**(*b*) Clamped corner—two sections.

The moment volume $V_M$ has been defined as

$$V_M \equiv \int (|M_1| + |M_2|)\, dA$$

where $M_1$, $M_2$ are the principal moments in the plate and dA is the element of plan area. For constant thickness plates and small amounts of reinforcement, such as generally will be the case for reinforced concrete plates, then $V_M$ is related to the volume of reinforcement ($V_R$) needed by the relation

$$V_M = V_R \,.\, \sigma \,.\, d$$

where $\sigma$ is the stress common to all the reinforcement.

The slab is required to support some specified loading, p, which will be assumed to be a uniform pressure, though any reasonable loading can be assumed.

The load will induce bending moments in the slab, $M_{\alpha\beta}$, which will vary

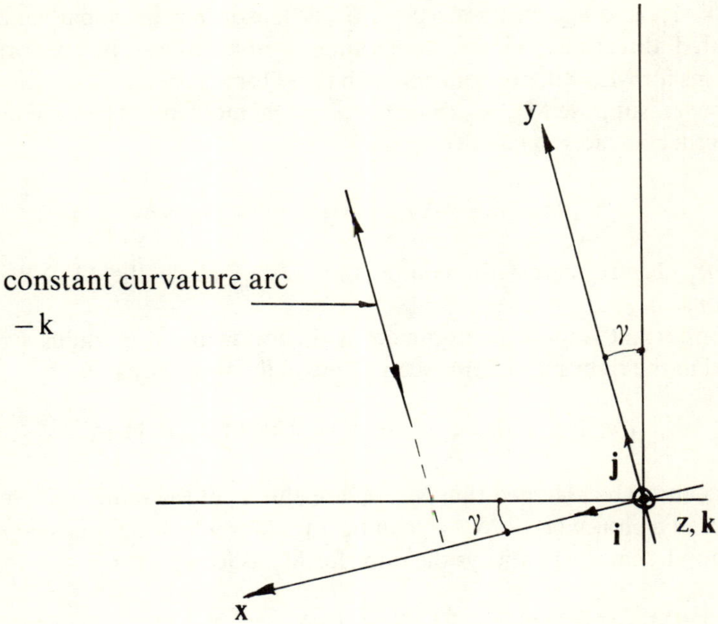


**Figure 5.7** General $-k$ surface—coordinate axes.

from point to point. As the plate deflects, curvatures ($\kappa_{\alpha\beta}$) will develop, associated with a *smooth* transverse displacement field $w = w(x, y)$. Here the expression $(\ )_{\alpha\beta}$ is used to mean the collection of all quantities $(\ )$. Thus for moments, $M_{\alpha\beta}$ is taken to mean $M_{xx}$, $M_{xy}$, $M_{yy}$, with $\alpha$ and $\beta$ being treated as free indices, and likewise for $\kappa_{\alpha\beta}$. Now, by virtual work, as the load p does work on the displacement w so internal work is absorbed by $M_{\alpha\beta}$ doing appropriate work on the components of $\kappa_{\alpha\beta}$. Symbolically this situation is described by

$$\int pw\,dA = \int M_{\alpha\beta} \cdot \kappa_{\alpha\beta} \cdot dA \tag{5.6.0}$$

where there is a summation on the repeated $\alpha$ and $\beta$ indices in the right-hand expression. Thus the right-hand side is the sum of terms obtained by letting $\alpha$ be first one coordinate, say x, and $\beta$ each coordinate (x, y) in turn; and then $\alpha$ is the second coordinate, y, and $\beta$ again takes on the values x and y. As the result, the moment work term in cartesian coordinates becomes

$$\int (M_{xx} \cdot \kappa_{xx} + M_{xy} \cdot \kappa_{xy} + M_{yx} \cdot \kappa_{yx} + M_{yy} \cdot \kappa_{yy})\,dA.$$

Each $M_{\alpha\beta}$ and $\kappa_{\alpha\beta}$, being a second order tensor, has principal values and associated directions. Unless something is done about it, the principal directions for $M_{\alpha\beta}$ will not coincide with those for $\kappa_{\alpha\beta}$.

However, suppose $M_1$, $M_2$ to be the *principal* moments. Then the moment work term contracts to just two terms

$$\int M_{\alpha\beta} \,.\, \kappa_{\alpha\beta} \,.\, dA = \int (M_1 \kappa_1 + M_1 \,.\, \kappa_2)\, dA,$$

where $\kappa_1$, $\kappa_2$ are merely the components of $\kappa_{\alpha\beta}$ along the $M_{1,2}$ principal directions.

An important step in the argument is the following. If modulus signs are inserted in the right-hand terms then an inequality results, as

$$\int M_{\alpha\beta} \,.\, \kappa_{\alpha\beta} \,.\, dA \leqslant \int (|M_1| \,.\, |\kappa_1| + |M_2| \,.\, |\kappa_2|)\, dA,$$

since it cannot be assumed that $M_\alpha$, $\kappa_\alpha$ are always of the same sign. Suppose however, if a choice (c) of $M_\alpha$, $\kappa_\alpha$ can be found for which the signs *do* always correspond, that is, are always the same for $M_\alpha$ as for $\kappa_\alpha$, then

$$\int (|M_1| \,.\, |\kappa_1| + |M_2| \,.\, |\kappa_2|)_c \, dA \leqslant \int (|M_1| \,.\, |\kappa_1| + |M_2| \,.\, |\kappa_2|)\, dA. \tag{5.6.1}$$

It should be emphasised that there is no special relation between $M_{\alpha\beta}$ and $\kappa_{\alpha\beta}$. The $M_{\alpha\beta}$ is an equilibrium system of bending moments and the $\kappa_{\alpha\beta}$ is a possible field of curvatures associated with a possible transverse displacement, w.

The essential minimal information derives from an interpretation of (5.6.1). First it is noted that for the minima, the $M_{\alpha\beta}$ and $\kappa_{\alpha\beta}$ must share principal directions and must correspond in sign.

To recover the expression for moment volume the curvatures are restricted to be constant or, should $M_\alpha = 0$, then $|\kappa_\alpha| < k$ allows $|k|$ to be taken out as a factor in (5.6.1) and we obtain

$$\int (|M_1| + |M_2|)_c \,.\, dA$$

$$\leqslant \int (\text{moment volume for any system not known to correspond})\, dA. \tag{5.6.2}$$

Hence the minimum is attained if associated principal curvature values can be found which correspond with the bending moments but have principal values which do not exceed the constant k value in absolute value.

If the layouts discussed earlier are checked against these criteria, it will be found that the criteria are met and hence the layouts and associated $V_M$'s are minimal.

In addition to these criteria for recognizing an optimal layout there is one important geometrical theorem which will now be proved. This is the theorem relating to the variation of the curvature transverse to a principal constant curvature arc.

Consider the most general context in which such an arc, assumed straight as seen in plan, can occur in an actual surface. The general motion of a constant curvature arc to describe a constant curvature surface can be represented by a radius vector **r** given by

$$\mathbf{r} = A(\gamma)\mathbf{i} + y\mathbf{j} + \left(a(\gamma) + b(\gamma)y - \frac{ky^2}{2}\right)\mathbf{k}. \tag{5.6.3}$$

Here A; a, b are kinematic degrees of freedom. The coordinates will be taken to be $\gamma$, y rather than x, y. Now, for convenience, the notation $d(\ )/d\gamma = (\ )_{,\gamma}$ will be adopted, and the following relations can be shown to hold:

$$\mathbf{i}_{,\gamma} = -\mathbf{j}, \mathbf{j}_{,\gamma} = \mathbf{i}, \mathbf{k}_{,\gamma} = 0$$

and

$$\mathbf{r}_{,\gamma} = (A_{,\gamma} + y)\mathbf{i} - A\,.\,\mathbf{j} + (a_{,y} + b_{,\gamma}\,.\,y)\mathbf{k}$$

$$\mathbf{r}_{,y} = \mathbf{j} + (b - ky)\mathbf{k}.$$

The surface slope is a vector quantity, and hence it is described by *two* components. Remember that these slopes are small, since the deflected surface is shallow. Then, for example, the slope in the y-direction along the $-k$ span is

$$S_y = b - ky.$$

Shallowness means that $kL \ll 1$, where L is a typical span in the problem. Also

$$A = O(L); \quad a = O(kL^2), \quad b = O(kL).$$

The quantities $\mathbf{r}_{,\gamma}$ and $\mathbf{r}_{,y}$ are tangent vectors to the surface along the $\gamma$ and y increasing directions respectively. They are not mutually orthogonal. However after some manipulation the *unit normal*, **n** can be evaluated to be

$$\mathbf{n} = -S_t\mathbf{i} - S_y\mathbf{j} + \mathbf{k} \tag{5.6.4}$$

where $S_t = [A\,.\,b + a_{,\gamma} + (b_{,\gamma} - kA)y]/[A_{,\gamma} + y]$ and $S_y = b - ky$.

The expression for $S_t$ then provides the result that if $S_t$ is the same for two distinct values of y it is the same for *all* values. In this case $S_t$ is given by

$$S_t = b_{,\gamma} - kA, \quad \text{since} \quad A_{,\gamma}(b_{,\gamma} - kA) = A\,.\,b + a_{,\gamma}.$$

When this is so the constant curvature direction is principal. A further ramification of such constant curvature surfaces is that the second principal curvature ($\kappa_2$, since $\kappa_1 = -k$ in this case), is given by

$$\kappa_2 = \frac{(b,_{\gamma\gamma} + b)}{A,_{\gamma} + y} - k. \tag{5.6.5}$$

This result shows that $\kappa_2$ is monotonically varying and once again this knowledge proves to be very valuable in studying optimal layouts where the condition $|\kappa_2| < k$ must be satisfied.

## 5.7 Conclusions

This chapter has been included to indicate to the reader some of the scope of geometry and statics applied to plate problems. In one sense the subject matter is out of place alongside the earlier chapters, because optimal plates have not yet entirely been accepted as practical realities.

However, the reader is urged to regard the content of this chapter as being of a nature of an investment in method, one which may or may not provide dividends. The point, nevertheless is clear that with a minimum of outlay, a method for finding a whole range of closed form solutions to anisotropic plate problems can be built up and used for exploration.

## 5.8 Exercises

Because of lack of space in the chapter proper a number of results are here suggested to the reader as exercises.

(1) Show that the simply supported square plate (side L) with pressure loading (p) requires a minimum moment volume of $V = 5pL^4/96$.

Further show that the central (neutral) zone absorbs 40% of this V if reinforced in a diagonal direction across a span of $L/\sqrt{2}$. The corner zones absorb 5% each, and edge reinforcement along the neutral zone boundary the remaining 40%.

(2) For the clamped square plate (side L) with pressure loading (p) show that the minimum moment volume requirement is $V = 13pL^4/768$.

(3) The equilateral triangular plate (side L) when pressure loaded requires a V of $0.00675pL^4$; the clamped equivalent requires a V of $0.00242pL^4$. Establish these results.

(It is suggested in all cases that maximum use be made of the corner requirements worked out, or quoted earlier).

(4) Show that the circular plate, when simply supported and loaded with a uniform pressure (p) requires a V of $\pi pR^4/4$. For the clamped edge circular plate and pressure loading show that $V = 7\pi pR^4/96$.

(5) Perhaps a more rational way of comparing these expressions is to express the quantity $pL^4$ or $pR^4$ as $pA^2$ where A is the plate plan area. When this is done then again the circular plate emerges as the *weakest* plate for given moment volume and either simply supported or clamped edge.

The comparative values are as in Table 5.0.

**Table 5.0** Values of moment volume (all $\times pA^2$)

| *Shape* | *Simply supported edge* | *Clamped edge* |
|---|---|---|
| Circular | 0.0798 | 0.0232 |
| Square | 0.0521 | 0.0170 |
| Equitriangular | 0.0350 | 0.0129 |

# 6 Bibliography and exercises

## 6.0 Bibliography

There is an extensive amount of material available on the subject of plate theory. Our problem here is one of selection. The first title to be mentioned should perhaps be S. P. Timoshenko's *Theory of Plates and Shells* (McGraw-Hill), first published in 1940, with a second edition (in collaboration with S. Woinowsky-Krieger) in 1959. As the title implies, shells as well as plates are dealt with, but the plates are static elastic plates only. The elastic stability of plates is dealt with in Timoshenko's companion work, *Theory of Elastic Stability* (McGraw-Hill), first published in 1936 and re-edited (in collaboration with J. M. Gere) in 1961. These are both standard works, rather than elementary textbooks.

Another standard work, which has exerted a very strong influence on many subsequent workers in this area of mechanics studies, is A. E. H. Love's *Theory of Elasticity* (Cambridge), first published in two volumes in 1892–3, with subsequent editions in one volume. The last (4th edition, 1927) is still in print and useful to some extent.

More elementary texts, which present plate topics at a level approximately equivalent to that aimed at in this text, include L. G. Jaeger, *Elementary Theory of Elastic Plates* (Pergamon, 1964); C. R. Calladine, *Engineering Plasticity* (Pergamon, 1969); and R. P. Johnson, *Structural Concrete* (McGraw-Hill, 1967), dealing respectively with elastic, plastic and yield line analysis of plates. All these books provide some exercises for the reader and give answers, but Imperial units may be encountered.

The treatment here has given some priority to the geometry of surfaces. The methods and results discussed in chapter 1 are of use beyond plate theory. Other more comprehensive accounts of surface geometry which use the methods of the Appendix are D. E. Rutherford, *Vector Methods* (Oliver and Boyd, 1939 and subsequent editions); K. L. Wardle, *Differential Geometry* (Routledge and Kegan Paul, 1965); C. E. Weatherburn, *Differential Geometry of Three Dimensions* (Cambridge, 2 vols, 1927, 1930) and T. J. Willmore, *Differential Geometry* (Oxford, 1959).

The Weatherburn volumes, though possibly available only in libraries, are especially useful to the engineer, although all the books were written with the mathematics student in mind. The Rutherford text can also very usefully serve as an aid to the study of the topics in chapter 1.

The contents of chapter 2 can be supplemented by reading books with a strength of materials flavour, for example, H. Ford, *Advanced Strength of Materials* (Longman, 1963) or S. C. Hunter, *Mechanics of Continuous Media* (Ellis Horwood, 1976). An aspect of chapter 2 is the choice of coordinates and expressions for the force and displacement quantities in the chosen system. There may be a need to consider the plate equations from the point of view of tensor formulations. The standard reference is A. E. Green and W. Zerna, *Theoretical Elasticity* (Oxford, 1954, 1968). This is a difficult book for engineers, and an alternative is B. Spain, *Tensor Calculus* (Oliver and Boyd, 1953), which takes a more leisurely route through the relevant tensor analysis.

Chapter 3 deals with the classical theory of elastic plates, and the amount of material available for collateral or extended reading is very large. The following titles are suggested, partly to draw the reader's attention to older, now less well-known but thoroughly readable accounts. For example, J. Prescott, *Applied Elasticity* (Longmans, 1924, and Dover), is still well worth taking down from the library shelf. So too is E. E. Sechler, *Elasticity in Engineering* (Wiley, 1952). Another out-of-print book, with a good coverage of circular elastic plates is C. E. Turner, *Introduction to Plate and Shell Theory* (Longmans, 1965). A comprehensive modern book, with some coverage of most relevant topics in elastic plates is R. Szilard, *Plate Theory—Classical and Numerical* (Prentice-Hall, 1974). Another recent book is L. H. Donnell, *Beams, Plates and Shells*, (McGraw-Hill, 1976). Anisotropic plates are dealt with in S. G. Leknitskii, *Anisotropic Plates* (Gordon and Breach, 1968).

Possible references to plastic plates (chapter 4) are also very numerous. Here the distinction must be drawn between plastic plates thought of as isotropic and homogeneous—therefore, in practice, solid metal plates, and nonhomogeneous plates such as reinforced concrete slabs. Each type has associated references. Homogeneous plastic plates are dealt with in P. G. Hodge, *Limit Analysis of Rotationally Symmetric Plates and Shells* (Prentice-Hall, 1963) and the same author's *Plastic Analysis of Structures* (McGraw-Hill, 1959). One of the founders of much of the relevant theory, W. Prager, can be read in his *Introduction to Plasticity* (Addison-Wesley, 1959). Calladine's *Engineering Plasticity* mentioned above can also be usefully read. M. A. Save and C. E. Massonet, *Plastic Analysis and Design of Plates, Shells and Disks* (North-Holland, 1972) is also useful.

The field of non-homogeneous plates and concrete slabs is well provided with references. An interesting work is R. H. Wood, *Plastic and Elastic Design of Slabs and Plates* (Thames and Hudson, 1961). A later work by L. L. Jones and R. H. Wood, *Yield-line Analysis of Slabs* (Thames and Hudson, 1967) is also worth consulting. A pioneering work is K. W. Johansen, *Yield-line Theory* (Cement and Concrete Association, 1962) an English translation of the 1943 Danish work. A recent work is R. Park and W. L. Gamble, *Reinforced Concrete Slabs* (Wiley, 1980).

Brief but readable accounts of yield-line and strip theory are contained in B. P. Hughes, *Limit State Theory for Reinforced Concrete Design* (Pitman, 1971, 1976) and F. K. Kong and R. H. Evans, *Reinforced and Prestressed Concrete* (Nelson 1975, 1980). The strip theory philosophy is given in English translation from the Swedish original in A. Hillerborg, *The Strip Method of Design* (Viewpoint, 1975).

The subject of optimum plates (chapter 5) is not well represented in book form and for further reading there is still a need to refer to the periodical literature. A convenient single reference is G. I. N. Rozvany, *Optimal Design of Flexural Systems* (Pergamon, 1976), and the list of references included there. A tabulation of moment volume information is contained in *Am. Soc. Civil Eng., Proc.* **99** (EM6) Dec. 1973, p. 1301. Some material in R. H. Wood, *Plastic and Elastic Slabs and Plates* (Thames and Hudson, 1961) is also relevant.

In addition to works in English there are relevant works in other languages. Just one will be noted, A. Sawczuk and T. Jaeger, *Grenztragfahigkeits Theorie der Platten* (Springer, 1963). This is a comprehensive and balanced account of plastic plates from the various viewpoints, and includes a discussion of supporting experimental work.

## 6.1 Exercises

Here are gathered exercises, some of which relate to topics which could not be included in the relevant chapters through lack of space. They may relate to the content of several chapters. All are intended to develop and test the reader's skills in formulation and problem solution.

(1) To begin this section, what is perhaps the first quantitative solution to a plate type problem will be considered. This example was posed and solved by John Wallis in the third part of his *Mechanica*, published in London in 1671. Even today the problem has some appeal and relevance.

As posed the problem relates to the use of a grid work of (wooden) beams, suitably jointed to fit together and span a square (or rectangular) area which is greater than the span of any of the available timbers.

Consider first a single module of the whole construction (Fig. 6.0). Let the end reactions be denoted by A, B, C, D, and the interactions by a, b, c, d.

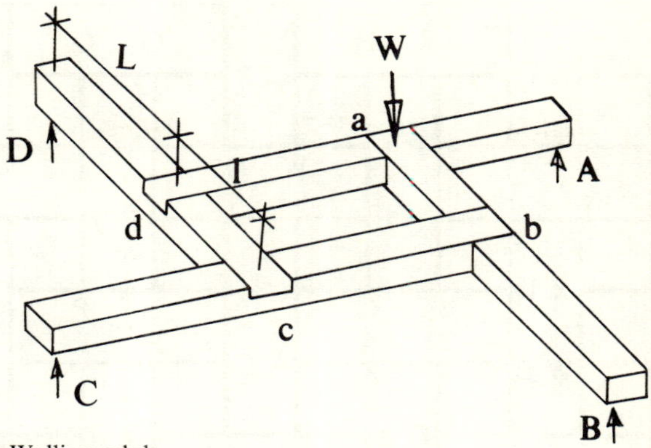


**Figure 6.0** The Wallis module.

When loaded with W as shown, calculate the values of these eight unknowns. The point to note is that the system is determinate, and hence statics alone will allow a solution to be found.

Show that

$$d\left[4+6\left(\frac{l}{L}\right)+4\left(\frac{l}{L}\right)^3+\left(\frac{l}{L}\right)^4\right]=W.\frac{L}{l}\left[1+\frac{l}{L}\right]^3.$$

When L/l = 2, for example, deduce that

$$a = 0.246W, \quad b = 0.369W, \quad c = 0.554W \quad \text{and} \quad d = 0.831W$$

and finally,

$$D = 0.277W, \quad \text{and} \quad A = 0.415W, \quad B = 0.123W, \quad C = 0.185W.$$

The Wallis problem is a system composed of an assembly of similar modules (Fig. 6.1). The whole area is a square of side 11L when the main member is of length 3L. There are 40 main members of length 3L, and 20 of length 2L to complete the edge details of the system. Suppose the self-weight of a member of length 3L to be T. Then this is the unit of loading to be assumed in the analysis. The total weight is then $53\frac{1}{3}$T. The system is assumed to be jointed so that a simple force of interaction is developed at each joint.

By an examination of the layout it can be concluded that the system is statically determinate. With the joint labelling as shown and using symmetry, then the total of 30 unknown forces can be shown to be given by the table of values below.

This problem can be used in various ways. A first procedure might be to check, essentially by back substituting into any relevant static equations, that the results are correct.

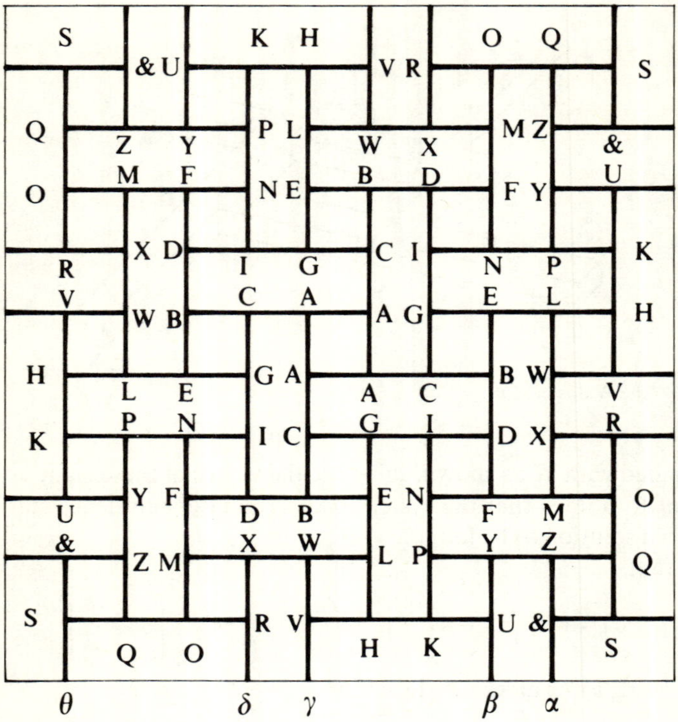


**Figure 6.1** Wallis's problem.

Wallis in his 1671 published solution solved for the 25 unknowns (excluding the edge reactions) exactly. If a common denominator of 680 334 is assumed, then all the 25 internal unknowns can be expressed in integral form (Table 6.0). For example, L = 2 513 665/680 334.

**Table 6.0** Wallis's problem: solutions

| | | | | | |
|---|---|---|---|---|---|
| A | 9.080 | L | 3.695 | W | 3.356 |
| B | 8.580 | M | 3.533 | X | 2.871 |
| C | 7.580 | N | 7.111 | Y | 3.212 |
| D | 7.185 | O | 5.572 | Z | 1.520 |
| E | 8.615 | P | 3.148 | & | 2.373 |
| F | 8.150 | Q | 2.584 | $\alpha$ | 1.520 |
| G | 7.475 | R | 5.076 | $\beta$ | 3.212 |
| H | 5.835 | S | 4.080 | $\gamma$ | 3.357 |
| I | 6.290 | V | 6.046 | $\delta$ | 2.871 |
| K | 4.969 | U | 5.758 | $\theta$ | 2.373 |

(2) Approximate solutions can be of various types. The following is an *elastic* deflection problem for which it is possible to establish upper and lower bounds, by elementary means.

Consider the deflection function

$$w = a(x^2 - 3y^2)^2(x - h)^2.$$

Investigate how the axes are arranged for this expression to describe an elastic equilateral triangular plate with fixed edges. Assume the plate is loaded with a uniform pressure p. Hence show that the maximum displacement $w_0$ lies between the bounds

$$228 < \frac{w_0 \, . \, 10^6 \, . \, D}{ph^4} < 344.$$

(3) As has been seen in chapter 4, there are quite a number of exact solutions known for simply-supported plastic plates governed by the square yield locus. There are, in contrast, fewer for clamped-edge plastic plates. A simple one is that for the clamped circular plate. An important, but complicated solution, is that for the square clamped plate, derived by E. N. Fox (*Phil. Trans. Roy. Soc.* (*A*), **277** (1974) 121–155). The main conclusion of this paper is that the collapse pressure, $p_c$, is given by $p_c \, . \, L^2/M_p = 42.851$. By reference to the data in that paper, and suitable interpolations, deduce that $A = 3.750$, $B = 7.074$ where A is the area interior to a square of side length 2 which spans the mechanism which develops, and B is the boundary length of this area

Hence show that

$$\frac{p_c}{M_p}\left(\frac{A}{B}\right)^2 = 3.01$$

for this case.

Though no proof has been offered for the proposition that $p_c A^2/M_p B^2$ should have a lower limit of 3, this result is important additional evidence in support of the inequality $p_c A^2/M_p B^2 \geqslant 3$ for clamped edge plates.

(4) Consider a six-sided simply supported plate with two axes of symmetry such that two opposite sides are parallel and of length b. The remaining four sides are of length a, are at right angles in pairs and form ends of the plate. To check your visualization, when $b = 0$, the shape becomes a square of side a. Investigate the values of $p_u \, . \, A/M_p$, $B^2/A$ and $p_u/M_p \, . \, (A/B)^2$ for a mechanism of conical type, as a function of the b/a ratio. Show that all three parameters have minima at the common value of $b/a = 2 - \sqrt{2}$. Show further that the minima are 21.94, 14.60, 1.50, and hence that these are not less than the corresponding values for the circular plate of 18.85, 12.56, 1.50. Note further that when $b/a = 2 - \sqrt{2}$, then the dihedral (maximum) slopes of the slab segments in the mechanism are all equal. Although it will not be proved, this

example is of special interest because, though not a fully symmetric shape of plate, $p_u = p_c$, that is the upper bound solution is exact.

(5) Consider a simply supported isotropic plate of arbitrary shape and pressure loaded to collapse. Show that if the mechanism of collapse is pyramid-shaped and has constant slope at the edge, then

$$\frac{p_c}{M_p}\left[\frac{A}{B}\right]^2 = 1.5$$

where A = area, B = perimeter length.

A sketch of a proof is

$$A = \int \frac{\Delta}{\phi}\frac{dB}{2} = \frac{\Delta}{\phi}\frac{B}{2}$$

where $\Delta$ = apex deflection, $\phi$ = edge slope, since $\phi$ is constant. Hence, the work equation gives $1/3 pA\Delta = mB\phi$ and the result follows. Note that a number of results, in chapter 4 especially, fit this specification.

(6) Space limitations have not permitted inclusion of a discussion of plate dynamics. Hence, as an exercise, confirm that

$$D\nabla^4 w = -m\ddot{w}, \quad (\ddot{.}) = \frac{\delta(..)}{\delta t}$$

is the relevant dynamic equation, where m is the unit mass of the plate. This equation can be solved, but only with more difficulty than the biharmonic. An approximate procedure which is useful, for example, in exploration of solutions to such equations is the *collocation method*. Essentially, the method aims to satisfy the equation *exactly* at a few chosen points, and not elsewhere, starting from an assumed shape for w, with suitable parameters. Suppose the natural frequency of a square, uniform, simply supported plate of side length a is sought. First assume $w(x, y, t) = A(x, y). \sin \omega_n t$, where $\omega_n$ is the frequency sought. Then assume say

$$A(x, y) = A\left[\left(\frac{a}{2}\right)^2 - x^2\right]^2\left[\left(\frac{a}{2}\right)^2 - y^2\right]^2$$

as a trial shape satisfying the boundary conditions, and just a single unknown parameter A, the amplitude. Hence, show that

$$\frac{m \,.\, a^4 \,.\, \omega_n^2}{D} = 35.7^2$$

is an estimate for $\omega_n$, if the dynamic equation is satisfied at the plate centre.

(7) Show that the information contained in the matrix equation (3.13) can be written alternatively as

$$M_{\alpha\beta} = D[(1-\nu)\kappa_{\alpha\beta} + \nu \,.\, \delta_{\alpha\beta} \,.\, \kappa_{\gamma\gamma}], \quad (\alpha, \beta = x, y).$$

Here the index notation and repeated indices (1.5) are being used, and $\delta$ is the delta symbol, such that $\delta_{\alpha\beta} = 1, \alpha = \beta; \delta_{\alpha\beta} = 0, \alpha \neq \beta$.

(8) Invert (3.1.3) and then show that $\kappa_{\alpha\beta}$ can be written in terms of $M_{\alpha\beta}$ as

$$E \, . \, \kappa_{\alpha\beta} = (1+\nu)M_{\alpha\beta} - \nu \, . \, \delta_{\alpha\beta} \, . \, M_{\gamma\gamma}, \quad E = Et^3/12.$$

(9) The potential energy of bending in a plate element is given by

$$\tfrac{1}{2}M_{\alpha\beta} \, . \, \kappa_{\alpha\beta} \, . \, dA \quad (A = \text{plate area}).$$

Hence show that the total potential energy V is given by

$$V = \frac{1}{2}\int M_{\alpha\beta}\kappa_{\alpha\beta}\, dA - \int pw\, dA$$

where $M_{\alpha\beta}$ are in equilibrium with loads p, $\kappa_{\alpha\beta}$ are derived from w, and $M_{\alpha\beta} = f(\kappa_{\alpha\beta})$ as in question (7) above.

10. Use the expressions in question (9) to show that

$$V = \frac{D}{2}\int \{[w,_{11} + w,_{22}]^2 - 2(1-\nu)[w,_{11}w,_{22} - w,_{12}^2]\} \, . \, dA - \int pw\, dA.$$

Hence show that if

$$w = A \operatorname{Sin}\frac{\pi x}{a} \operatorname{Sin}\frac{\pi y}{a}$$

then the maximum displacement (A) for a pressure loaded square simply supported plate is obtained by minimizing V, when

$$A = \frac{4}{\pi^6} \cdot \frac{pa^4}{D}.$$

# Appendix: Geometry of surfaces

## A.0 The need for geometry

The design of a plate as a load-bearing element in a structure or machine requires that the *deformation* of the plate be understood. Initially a plate is plane, or nearly so. When loaded it deflects into some shape which is not plane; hence there is a need to describe this new, curved shape in order to compare with the unloaded plane shape and so evaluate the internal actions generated.

For the purposes of this chapter the plate will be characterized by the "middle surface" in the case of a solid plate, or some other representative surface for composite plates. The middle surface, as the name implies, is the central layer equidistant from the faces of the plate. Thus the deformation of the plate will be characterized by the deformation of the middle surface. This is reasonable, since most plates are thin compared with the distances spanned by the plate and, as already discussed for beams (1.6), the plate will be assumed to obey the Euler–Bernoulli hypothesis of plane sections remaining plane. However, unlike beams whose *geometry* of deformation is comparatively simple, plates are by definition two-dimensional objects and their deformation presents other aspects which can usefully be studied in this chapter.

## A.1 Geometry of a plane curve—curvature

A plane curve possesses *shape* by virtue of the curvature of the curve. Consider a small piece of the curve. Assuming this piece to be circular, of radius $\rho$ (see Fig. A.0), then the *curvature* of this portion of curve is defined as $\kappa \equiv 1/\rho$. The quantity $\rho$ is known as the *radius* of curvature.

If the element of arc length is denoted by ds and $d\psi$ is the angle between adjacent radii of curvature then

$$\kappa = 1/\rho = \frac{d\psi}{ds}. \qquad (A.1.0)$$

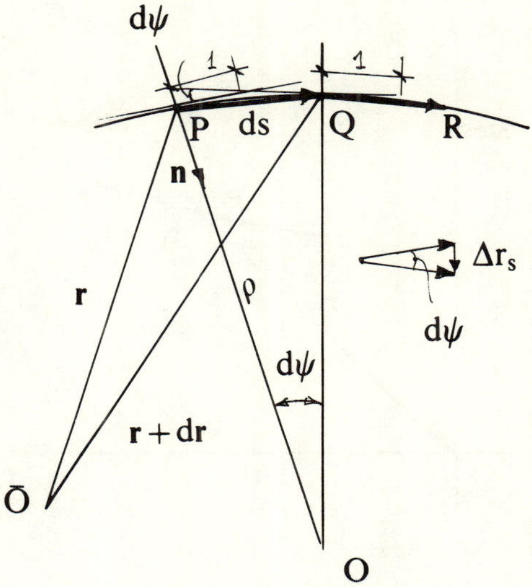


**Figure A.0** Geometry of a curve.

Now rework this definition in a vector form. Consider a radius vector $\mathbf{r}$ to point P. Then for Q, the radius vector will be $\mathbf{r}+d\mathbf{r}$. Hence $d\mathbf{r}$ is a *tangent* vector. Further, $d\mathbf{r}/ds$ is a *unit* tangent vector $(\mathbf{t})$ because $|d\mathbf{r}| = ds$. Denote $d\mathbf{r}/ds$ by $\mathbf{r}_s$, and consider what meaning can be attached to $\mathbf{r}_{ss}(=d^2\mathbf{r}/ds^2)$.

Now

$$\mathbf{r}_{ss} = \frac{(\mathbf{r}_s)_Q - (\mathbf{r}_s)_P}{ds} = \frac{\Delta\mathbf{r}_s}{ds}. \tag{A.1.1}$$

From Fig. A.0 we see that $(\mathbf{r}_s)_Q$ and $(\mathbf{r}_s)_P$ are inclined at $d\psi$ and the difference is directed along the inward normal $\mathbf{n}$. Comparing (A.1.0) and (A.1.1) it is seen that

$$\kappa = \mathbf{n}\,.\,\mathbf{r}_{ss}, \tag{A.1.2}$$

where $\mathbf{n}$ is a *unit* inward normal, *directed toward the centre of curvature, O.* This is a convenient vectorial expression for curvature of a plane curve which will also be found useful in the later discussion of surfaces.

*Example:* Derive an expression for the curvature $(\kappa)$ of a plane curve in terms of cartesian coordinates (Fig. A.1).

Now

$$\mathbf{r} = x\mathbf{i} + y\mathbf{j},$$

where $\mathbf{i}, \mathbf{j}$ are fixed unit vectors. The curve is a one-parameter object such that $y = f(x)$, say, where $f(x)$ is given.

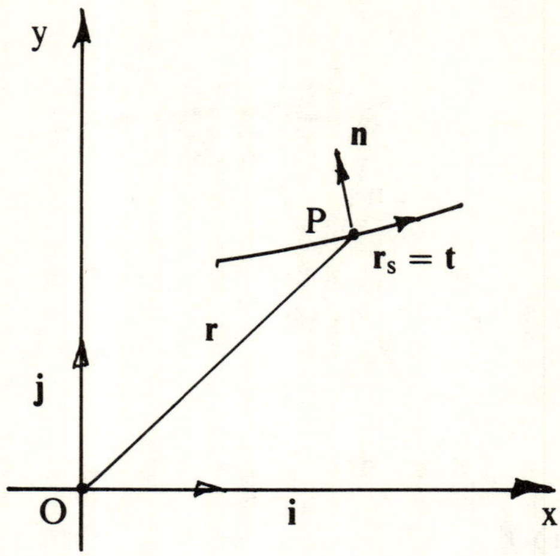


**Figure A.1** Plane curve—curvature.

Hence

$$\mathbf{r} = x\mathbf{i} + f(x)\mathbf{j}.$$

Now

$$ds^2 = dx^2 + dy^2 = (1 + f_x^2)\,dx^2,$$

$$\text{where} \quad f_x = \frac{df}{dx}, \qquad \text{or} \qquad ds = \sqrt{(1+f_x^2)}\,dx.$$

$$\therefore \quad \mathbf{r}_s = \frac{d\mathbf{r}}{ds} = \frac{d\mathbf{r}}{dx}\cdot\frac{dx}{ds} = [\mathbf{i} + f_x\mathbf{j}]\,.\frac{1}{\sqrt{1+f_x^2}}, = \text{unit tangent vector.}$$

$$\therefore \quad \mathbf{n} = \textbf{unit normal vector} = \frac{-f_x\mathbf{i} + \mathbf{j}}{\sqrt{1+f_x^2}}$$

and note that $\mathbf{n}\,.\,\mathbf{t} = \mathbf{n}\,.\,\mathbf{r}_s = 0$, since $\mathbf{n}$ is the normal and $\mathbf{r}_s$ the tangent to the curve.

Then

$$\mathbf{r}_{ss} = \frac{d\mathbf{t}}{ds} = \frac{d\mathbf{t}}{dx}\cdot\frac{dx}{ds} = \frac{(f_x\mathbf{i} + \mathbf{j})f_{xx}}{(1+f_x^2)^2}.$$

Hence

$$\kappa \equiv \mathbf{n}\,.\,\mathbf{r}_{ss} = \frac{f_{xx}}{(1+f_x^2)^{3/2}} = \frac{d^2y/dx^2}{\left[1 + \left(\dfrac{dy}{dx}\right)^2\right]^{3/2}}.$$

## A.2 Length measurement on a surface—first fundamental form

Before any useful analysis of surface geometry can be undertaken, certain preliminaries must be gone through. The first of these is to discuss the measurement of lengths on a surface. Consider the surface to be covered with a coordinate system of two independent variables u, v. There is no need to be specific about u and v at the moment, except to note that each point on the surface can be identified by a (u, v) pair of coordinates. Consider a typical point P on the surface (Fig. A.2), and a neighbouring point Q. Choose a fixed origin, O, and let the point P be located by the radius vector ($\mathbf{r}$) from O such that

$$\mathbf{r} = \mathbf{r}(u, v) \tag{A.2.0}$$

Distances measured along the surface, say from P to Q, will be denoted by arc length s, or in this particular case by ds since the points P, Q are regarded as infinitesimally separated. From the vector relations indicated on Fig. A.2 it is seen that ds is the modulus of $d\mathbf{r}$. In order to evaluate ds the standard approach is to form

$$ds^2 = d\mathbf{r} \, . \, d\mathbf{r}. \tag{A.2.1}$$

But

$$d\mathbf{r} = \mathbf{r}_u \, du + \mathbf{r}_v \, dv \tag{A.2.2}$$

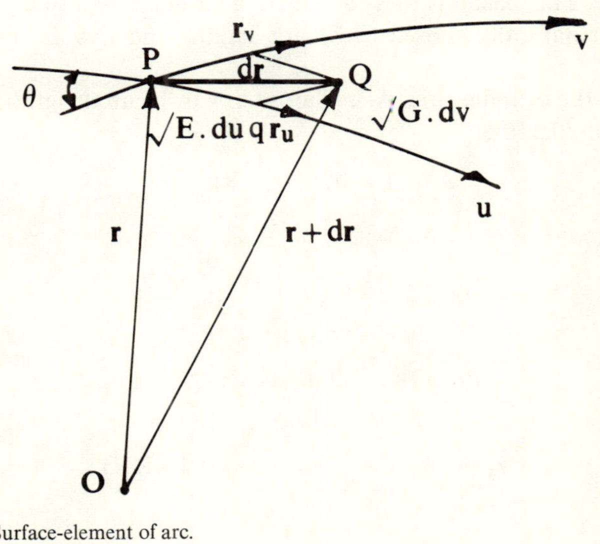


**Figure A.2** Surface-element of arc.

since $\mathbf{r} = \mathbf{r}(u, v)$, and where

$$\mathbf{r}_u \equiv \frac{d\mathbf{r}}{du}, \mathbf{r}_v \equiv \frac{d\mathbf{r}}{dv}.$$

Hence

$$\begin{aligned} ds^2 &= (\mathbf{r}_u\, du + \mathbf{r}_v\, dv) . (\mathbf{r}_u\, du + \mathbf{r}_v\, dv), \\ &\equiv E\, du^2 + 2F\, du\, dv + G\, dv^2, \qquad \text{(A.2.3)} \\ &\equiv \mathbf{I}. \end{aligned}$$

Here the following definitions have been made:

$$E \equiv \mathbf{r}_u . \mathbf{r}_u, \quad F \equiv \mathbf{r}_u . \mathbf{r}_v = \mathbf{r}_v . \mathbf{r}_u, \quad G \equiv \mathbf{r}_v . \mathbf{r}_v. \qquad \text{(A.2.4)}$$

The expression (A.2.3) is known as the *First Fundamental Form* (I), and E, F, G, as the *coefficients* of this form. These definitions and (A.2.3) are very important and can be interpreted as follows. Suppose $dv \equiv 0$. Then Q is at q and the change du produces an arc length Pq which is given by $ds = \sqrt{E}\, du$, from (A.2.3). These are scalar quantities. But $\mathbf{r}_u$ is a vector, and a *tangent* vector to the surface at P along the direction u increasing, namely P to q. Similarly $\mathbf{r}_v$ is a tangent vector to the surface at P but along the direction v increasing, as indicated in Fig. A.2.

This is not too surprising, but it is very important: to repeat, $\mathbf{r}_u$, $\mathbf{r}_v$ are *tangent* vectors along the u, v lines respectively. Hence $d\mathbf{r}$, obtained when both u and v vary to a point such as Q, is itself a tangent vector and relates to $\mathbf{r}_u$, $\mathbf{r}_v$ via (A.2.2). These tangent vectors however are often *not* unit vectors. Roughly speaking they are seldom unit vectors when the u, v lines are curved in any sense. The reason is that such lack of straightness usually means that u, v are not naturally chosen to be arc lengths and this destroys the unit quality.

Consider the cylinder, Fig. A.3. Choose u, v to be arc lengths around and along the cylinder, then

$$\mathbf{r} = 0 . \mathbf{i} + v\mathbf{j} + R\mathbf{k}$$

and

$$\frac{d\mathbf{i}}{d\theta} = -\mathbf{k}, \quad \frac{d\mathbf{k}}{d\theta} = \mathbf{i}, \quad \frac{d\mathbf{j}}{d\theta} = 0.$$

Hence

$$d\mathbf{r} = \overrightarrow{PQ} = R\, d\theta . \mathbf{i} + dv . \mathbf{j} + 0 . \mathbf{k}.$$

But

$$\begin{aligned} \mathbf{r}_u &= \frac{1}{R} . \mathbf{r}_\theta = \frac{1}{R} . [0 . -k + 0 . \mathbf{j} + R . \mathbf{i}], \\ &= \mathbf{i} \quad \text{and} \quad \mathbf{r}_v = \mathbf{j}. \end{aligned}$$

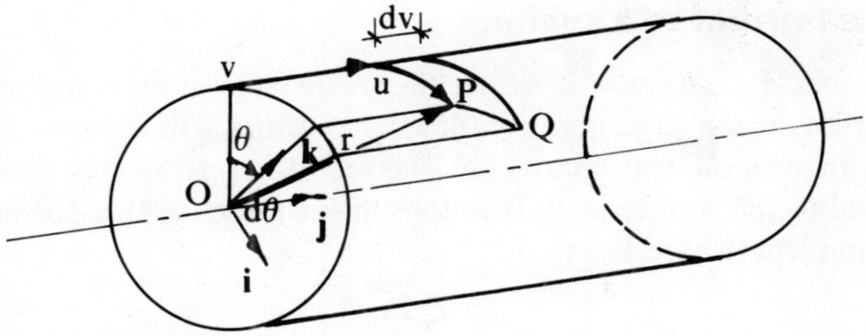


**Figure A.3** Geometry on the cylinder ($u = R\theta$, $du = R\,d\theta$).

Hence

$$E \equiv \mathbf{r}_u . \mathbf{r}_u = \mathbf{i}.\mathbf{i} = 1, \quad F \equiv \mathbf{r}_u . \mathbf{r}_v = \mathbf{i}.\mathbf{j} = 0, \quad G \equiv \mathbf{r}_v . \mathbf{r}_v = \mathbf{j}.\mathbf{j} = 1.$$

In this case, with this choice of coordinates, $E = G = 1$, $F = 0$. However, suppose $u = \theta$, with v, as before, the (axial) arc length (Fig. A.3). Then

$$\mathbf{r}_u = \mathbf{r}_\theta = R\mathbf{i}, \quad \mathbf{r}_v = \mathbf{j} \quad \text{(as before)}.$$

Hence

$$E = \mathbf{r}_u . \mathbf{r}_u = R^2, \quad F = 0, \quad G = 1.$$

Hence $E \neq G$, though $F = 0$ still.

Yet other coordinates might be chosen, such as a helix and an axial length. In such cases, E, F and G take on non-zero, and probably more complicated, values.

*Exercise:* Consider the surface $z = Kxy$, where x, y, z are cartesian axes (that is all three axes are straight and mutually perpendicular). If x, y are chosen to be the independent coordinates, show that the coefficients of the first fundamental form are $E = 1 + (Ky)^2$, $F = K^2xy$, $G = 1 + (Kx)^2$.

The surface $z = Kxy$ is known as a *hyperbolic paraboloid* (H.P.).

The discussion thus far allows quite general choice of u, v coordinates. In the examples above the coordinates were *orthogonal*, and then $F = 0$, since $\mathbf{r}_u$ and $\mathbf{r}_v$ are orthogonal. If $\phi$ is the angle between the u, v lines, then by definition of the various quantities,

$$\mathbf{r}_u . \mathbf{r}_v \equiv F = \sqrt{EG}\,.\cos\phi.$$

Hence

$$\sin\phi = \sqrt{1 - \cos^2\phi} = \sqrt{1 - \frac{F^2}{EG}} = \frac{H}{\sqrt{EG}}, \quad \text{where} \quad H^2 \equiv EG - F^2.$$

*Exercise:* For the H.P. show that $H^2 = 1 + K^2 . (x^2 + y^2)$.

## A.3 The normal to a surface

At every point P on a smooth surface there exist at least two tangent vectors, for example $\mathbf{r}_u$ and $\mathbf{r}_v$. The plane of these two vectors is the *tangent plane* at P in which all such tangent vectors lie. The vector *normal* to this plane at P is the normal to the surface at P. It is convenient to define this normal vector to be of unit length as follows:

$$\mathbf{n} \equiv \frac{\mathbf{r}_u \times \mathbf{r}_v}{|\mathbf{r}_u \times \mathbf{r}_v|}. \tag{A.3.0}$$

The denominator in this vector product expression is given by

$$\sqrt{EG}\sin\phi = H,$$

since $\mathbf{r}_u$, $\mathbf{r}_v$ are inclined at $\phi$ and $|\mathbf{r}_u| = \sqrt{E}$, $|\mathbf{r}_v| = \sqrt{G}$. Since the surface is smooth, every point on the surface possesses a unique normal, the only ambiguity being the *direction* in which $\mathbf{n}$ is considered positive. Once u and v are defined, then (A.3.0) selects the direction for $\mathbf{n}$ positive.

When adjacent points such as P and Q of Fig. A.2 are considered, each has an associated normal. Generally these normals will be oriented in non-parallel directions, though the vector difference between the two normals is infinitesimal by virtue of smoothness. Let $\mathbf{n}$ be the (unit) normal at P, then $\mathbf{n}+d\mathbf{n}$ can be considered to be the (unit) normal at Q. Since $\mathbf{n}$ and $\mathbf{n}+d\mathbf{n}$ are unit vectors, $d\mathbf{n}$ must be normal to $\mathbf{n}$ otherwise the length of $\mathbf{n}+d\mathbf{n}$ could not

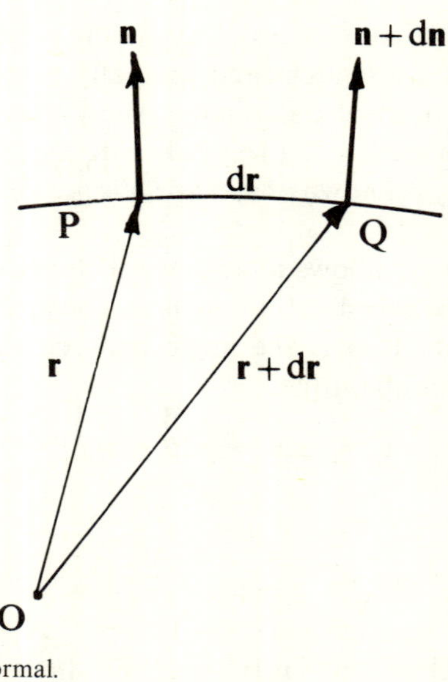


**Figure A.4** The surface normal.

be unity. Hence d**n** must lie in the *tangent plane* at P. This is an important conclusion.

Consider the following—what condition must hold if **n** and **n**+d**n** are to be coplanar? Generally these two vectors will be skew. For them to be coplanar then d**n** must lie in the plane of d**r** and **n**.

This latter condition requires that

$$\mathbf{n} \,.\, \mathrm{d}\mathbf{n} \times \mathrm{d}\mathbf{r} = 0. \qquad \text{(A.3.1)}$$

Taken a stage further, d**n** and d**r** must be parallel (co-axial) if this condition is to hold, since otherwise d**n** × d**r** will have a component in the **n** direction and the scalar (A.3.1) will not be zero. The next stage is to note that each of d**n** and d**r** is a linear expression in terms of du and dv. Hence (A.3.1) is a quadratic in du and dv. But a ratio of du/dv corresponds to a particular direction of Q in relation to P. Hence (A.3.1) supplies *two specific* directions through P—the *principal directions* at P—for which the normals at neighbouring points *are* coplanar. This topic will be returned to later when the importance of these directions can be further explored.

*Exercise:* Show that the normal to the hyperbolic paraboloid z = Kxy is

$$\mathbf{n} = \left(-\mathrm{Ky}\mathbf{i} - \mathrm{Kx}\mathbf{j} + \mathbf{k}/\sqrt{1 + \mathrm{K}^2\mathrm{x}^2 + \mathrm{K}^2\mathrm{y}^2}\right).$$

## A.4 Normal curvature—second fundamental form

The notion of curvature of a surface is an extension of the notion of curvature of a plane curve. For a plane curve a small piece of curve in the neighbourhood concerned is examined. It is assumed to be accurately described as locally circular and the associated curvature is then the inverse of the radius of this piece of circular curve. Hence *curvature* has the dimensions of $\mathrm{L}^{-1}$.

On a surface, the study is begun by first drawing the normal, **n**, at the point P of interest. But now the curvature is *direction* dependent.

A plane ($\gamma$) containing **n** and oriented so as to contain the required direction on the surface is drawn to cut the surface in a piece of (plane) curve Γ (Fig. A.5).

From what was discussed in A.3 there is no guarantee that **n**+d**n**, the unit normal at Q, will lie in $\gamma$, unless Γ is a principal direction. Now Γ is regarded as just another piece of plane curve and there will be an associated curvature ($\kappa_p$) at P. This is the curvature of the surface at P in the direction P to Q. As Q is chosen to be, in turn, different points in the neighbourhood of P, so $\kappa_p$ will vary. (How $\kappa_p$ varies is itself of considerable interest).

From what has already been described, d**n**, the change in **n** away from P, lies parallel to the tangent plane. This is merely a consequence of **n** at all points being restricted to being a *unit* normal. Any change must therefore

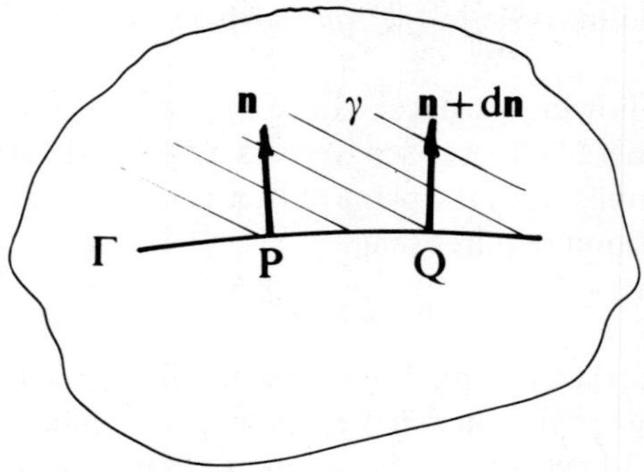


**Figure A.5** Surface curvature.

take place *normal* (transverse) to the existing normal **n**. This point will be pursued in the next section.

Here the aim is to look at some properties of the second derivatives of **r**. Now **n** is normal to each of $\mathbf{r}_u$ and $\mathbf{r}_v$. Hence $\mathbf{n}\,.\,\mathbf{r}_u = 0$. Thus it follows that, by taking $(\ )_u$ of this expression,

$$\mathbf{n}\,.\,\mathbf{r}_{uu} + \mathbf{n}_u\,.\,\mathbf{r}_u = 0,$$

where

$$(\ )_u \equiv \frac{\partial(\ )}{\partial u}, \quad (\ )_{uu} \equiv \frac{\partial^2(\ )}{\partial u^2}.$$

Rewriting this expression, the normal component of $\mathbf{r}_{uu}$ will be denoted by L when

$$L \equiv \mathbf{n}\,.\,\mathbf{r}_{uu} = -\mathbf{n}_u\,.\,\mathbf{r}_u.$$

In a similar manner

$$N \equiv \mathbf{n}\,.\,\mathbf{r}_{vv} = \mathbf{n}_v\,.\,\mathbf{r}_v. \tag{A.4.0}$$

Finally

$$M \equiv \mathbf{n}\,.\,\mathbf{r}_{uv} = -\mathbf{n}_u\,.\,\mathbf{r}_v = -\mathbf{n}_v\,.\,\mathbf{r}_u.$$

These three components L, M, N are known as the *coefficients of the second fundamental form* (II).

The second fundamental form itself is defined as

$$II \equiv L\,du^2 + 2M\,du\,dv + N\,dv^2. \tag{A.4.1}$$

The relationship between **r**, **n** and II is then seen to be

$$\begin{aligned}-d\mathbf{n}\,.\,d\mathbf{r} &= -(\mathbf{n}_u\,du+\mathbf{n}_v\,dv)\,.\,(\mathbf{r}_u\,du+\mathbf{r}_v\,dv)\\ &= -(\mathbf{n}_u\,.\,\mathbf{r}_u\,du^2+(\mathbf{n}_u\,.\,\mathbf{r}_v+\mathbf{n}_v\,.\,\mathbf{r}_u)\,du\,dv+\mathbf{n}_v\,.\,\mathbf{r}_v\,dv^2),\\ &= L\,du^2+2M\,du\,dv+N\,dv^2,\\ &= II. \end{aligned} \tag{A.4.2}$$

*Exercise:* For the H. P., $z = Kxy$, show that

$$L = N = 0, \quad M = K/\sqrt{1+K^2(x^2+y^2)}.$$

By extension from the curvature definition for a plane curve, (A.1.2), the *normal curvature* of the surface can be defined to be

$$\kappa_n \equiv \mathbf{n}\,.\,\mathbf{r}_{ss}. \tag{A.4.3}$$

Now

$$\mathbf{r}_s = \mathbf{r}_u\,.\,u_s+\mathbf{r}_v\,.\,v_s,$$

and

$$\mathbf{r}_{ss} = \mathbf{r}_{uu}\,.\,u_s^2+2\,.\,\mathbf{r}_{uv}\,.\,u_s v_s+\mathbf{r}_{vv}\,.\,v_s^2+\mathbf{r}_u u_{ss}+\mathbf{r}_v v_{ss}.$$

Hence

$$\begin{aligned}\kappa_n = \mathbf{n}\,.\,\mathbf{r}_{ss} &= Lu_s^2+2M\,.\,u_s\,.\,v_s+N\,.\,v_s^2,\\ &= \frac{II}{I}.\end{aligned} \tag{A.4.4}$$

This is an important basic form for calculation of $\kappa_n$.

## A.5 The derivatives of n—the Weingarten equations

If d**n** is parallel to the tangent plane, so too are $\mathbf{n}_u$ and $\mathbf{n}_v$.

Hence

$$\mathbf{n}_u = a\mathbf{r}_u+b\mathbf{r}_v \tag{A.5.0}$$

where a, b are coefficients to be found.

Now

$$-L = \mathbf{n}_u\,.\,\mathbf{r}_u = a\mathbf{r}_u\,.\,\mathbf{r}_u+b\mathbf{r}_u\,.\,\mathbf{r}_v = Ea+Fb,$$

from the definition of L.

Again

$$-M = \mathbf{n}_u\,.\,\mathbf{r}_v = a\mathbf{r}_v\,.\,\mathbf{r}_u+b\mathbf{r}_v\,.\,\mathbf{r}_v = Fa+Gb.$$

Hence, here are two equations for a and b. Solve to obtain

$$\begin{bmatrix} a \\ b \end{bmatrix} = \frac{-1}{H^2}\begin{bmatrix} G & -F \\ -F & E \end{bmatrix}\begin{bmatrix} L \\ M \end{bmatrix}.$$

Hence

$$a = \frac{1}{H^2}[FM - GL], \quad b = \frac{1}{H^2}[FL - EM],$$

where $H^2 = EG - F^2$.

In a similar manner, $\mathbf{n}_v$ can be evaluated.

Finally

$$\mathbf{n}_u = \left(\frac{FM - GL}{H^2}\right)\mathbf{r}_u + \left(\frac{FL - EM}{H^2}\right)\mathbf{r}_v,$$

$$\mathbf{n}_v = \left(\frac{FN - GM}{H^2}\right)\mathbf{r}_u + \left(\frac{FM - EN}{H^2}\right)\mathbf{r}_v. \qquad \text{(A.5.1)}$$

These two equations are known as the Weingarten equations. They will be used extensively in the discussion of curvature to follow.

## A.6 Directions on a surface

Consider a typical point, P, on a smooth surface (Fig. A.6), and the u, v coordinate curves through this point. Suppose there are two directions, 1, 2 through P to be investigated. Let $PQ_1$ be denoted by $d\mathbf{r}$ and $PQ_2$ by $\delta\mathbf{r}$.

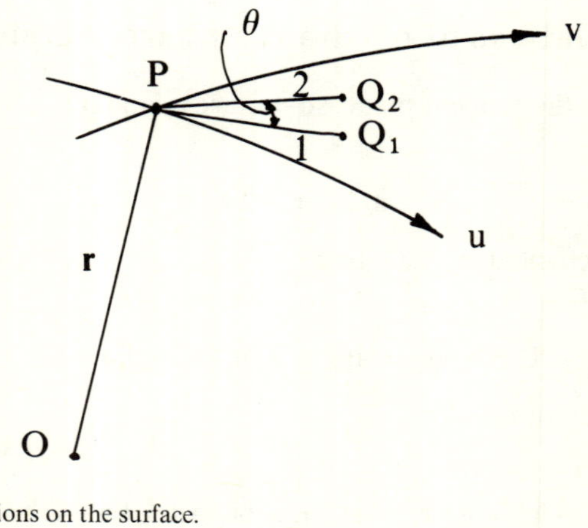


**Figure A.6** Directions on the surface.

Then

$$\mathbf{dr} = \mathbf{r}_u \,.\, du + \mathbf{r}_v \, dv,$$
$$\delta\mathbf{r} = \mathbf{r}_u \delta u + \mathbf{r}_v \delta v.$$

Hence

$$\mathbf{dr} \,.\, \delta\mathbf{r} = E\, du \,.\, \delta u + F(du \,.\, \delta v + dv \,.\, \delta u) + G\, dv \,.\, \delta v.$$

If the angle between these two directions is $\theta$, then $ds \,.\, \delta s \cos\theta = \mathbf{dr} \,.\, \delta\mathbf{r}$ and the two directions will be *orthogonal* if $\cos\theta = 0$. In this case

$$E\left(\frac{du}{dv}\cdot\frac{\delta u}{\delta v}\right) + F\left(\frac{du}{dv} + \frac{\delta u}{\delta v}\right) + G = 0. \tag{A.6.0}$$

Now the quantities $du/dv$, $\delta u/\delta v$ essentially describe directions and hence (A.6.0) is the condition that two such directions should be orthogonal.

*Exercises:* Show that the angle ($\alpha$) between the tangent vectors $\mathbf{r}_x$, $\mathbf{r}_y$ for the H.P. is in general not $\pi/2$. Prove that $\operatorname{Tan}\alpha = H/K^2xy$, where

$$H^2 = 1 + K^2(x^2 + y^2).$$

## A.7 The principal curvatures

Earlier, in (A.2), it was shown that in general the normals drawn at adjacent points on a smooth surface are non-coplanar. A condition was proposed as a test for the situation when they *are* coplanar. This condition is that $\mathbf{r}$ and $\mathbf{n}$ are such that

$$\mathbf{n} \,.\, d\mathbf{n} \times \mathbf{dr} = 0. \tag{A.7.0}$$

It can be concluded that for (A.7.0) to hold then d**n** and d**r** must be parallel. If not, their vector product will be non-zero and this will point in the **n** direction since d**n** and d**r** lie in the tangent plane. Now, if d**n** and d**r** are parallel, then

$$d\mathbf{n} = \text{scalar} \,.\, \mathbf{dr}. \tag{A.7.1}$$

Since **n** is dimensionless, being a direction, and **r** has dimensions of length, being a *position* vector, then the scalar must have dimensions of $L^{-1}$. Indeed, the scalar is in the nature of a curvature. It will now be shown that it is the surface curvature in the direction d**r**, which in this case is a principal curvature. Calling this curvature $\kappa_p$, then

$$\kappa_p \,.\, \mathbf{dr} + d\mathbf{n} = 0. \tag{A.7.2}$$

This equation is known as *Rodrigues' formula.*

Now from (A.4.4)

$$\kappa_n = \frac{II}{I} = \frac{-d\mathbf{n}\,.\,d\mathbf{r}}{ds^2}.$$

By taking d**r** of (A.7.2) it is seen that

$$\kappa_p\, d\mathbf{r}\,.\,d\mathbf{r}\ (= \kappa_p\, ds^2)$$
$$= -d\mathbf{n}\,.\,d\mathbf{r}.$$

Hence $\kappa_p = \kappa_n$, in this case.
But

$$d\mathbf{r} = \mathbf{r}_u\, du + \mathbf{r}_v\, dv,$$

and

$$d\mathbf{n} = \mathbf{n}_u\, dn + \mathbf{n}_v\, dv.$$

Also from the Weingarten equations

$$\mathbf{n}_u\,.\,\mathbf{r}_u = \left(\frac{FM-GL}{H^2}\right)E + \left(\frac{FL-EM}{H^2}\right)F = -L.$$

and

$$\mathbf{n}_v\,.\,\mathbf{r}_u = \left(\frac{FN-GM}{H^2}\right)E + \left(\frac{EM-EN}{H^2}\right)F,$$
$$= \frac{EFN-EGM+F^2M-EFN}{EG-F^2} = -M.$$

Now take $\mathbf{r}_u$ of (A.7.2) when

$$(\kappa_p E - L)\,du + (\kappa_p\,.\,F - M)\,dv = 0.$$

Further, take $\mathbf{r}_v$ of (A.7.2) when

$$(\kappa_p F - M)\,du + (\kappa_p\,.\,G - N)\,dv = 0.$$

These two homogeneous equations in (du, dv) imply that

$$\begin{vmatrix} (\kappa_p\,.\,E - L) & (\kappa_p\,.\,F - M) \\ (\kappa_p\,.\,F - M) & (\kappa_p\,.\,G - N) \end{vmatrix} = 0,$$

or

$$\boxed{H^2\kappa_p^2 - \kappa_p(EN - 2FM + GL) + T^2 = 0} \qquad \text{(A.7.3)}$$

where

$$H^2 = EG - F^2, \quad T^2 = LN - M^2.$$

The two roots, $\kappa_p$, provided by (A.7.3) are the *two principal curvatures*.
*Exercise:* For the H. P., z = Kxy, show that the principal curvatures are given by $\kappa_1 = K/H^3$, $\kappa_2 = -K/H$.

## A.8 Principal directions

Along any general direction du/dv the (normal) curvature $\kappa_n$ is given by

$$\kappa_n = \frac{\mathrm{II}}{\mathrm{I}} = \frac{L\,du^2 + 2M\,du\,dv + N\,dv^2}{E\,du^2 + 2F\,du\,dv + G\,dv^2},$$

$$= \frac{L.\lambda^2 + 2M\lambda + N}{E\lambda^2 + 2F\lambda + G}. \tag{A.8.0}$$

This curvature has turning values of $\kappa_n$ when $\lambda(= du/dv)$ takes on the values implied by

$$\frac{d\kappa_n}{d\lambda} = 0 = \frac{(E\lambda^2 + 2F\lambda + N).(2L\lambda + 2M) - (L\lambda^2 + 2M\lambda + N).(2E\lambda + 2F)}{(E\lambda^2 + 2F\lambda + G)^2},$$

i.e. when the numerator is zero. Hence

$$(FL - EM)\lambda^2 + (GL - EN)\lambda + (GM - FN) = 0. \tag{A.8.1}$$

The roots $\lambda_1$, $\lambda_2$ of this equation define two directions through point P for which the curvature has extreme values.

Now from (A.8.1)

$$\lambda_1 + \lambda_2 = -\frac{(GL - EN)}{(FL - EM)},$$

and

$$\lambda_1 . \lambda_2 = \frac{GM - FN}{FL - EM}. \tag{A.8.2}$$

If these roots for the directions of extremal $\kappa_n$ are substituted into (A.6.0) it is found that

$$E\left(\frac{GM - FN}{FL - EM}\right) - F\left(\frac{GL - EN}{FL - EM}\right) + G = 0.$$

This is an identity. Hence it is concluded that the extremal $\kappa_n$ directions are *orthogonal.*

It remains now only to identify these extreme values $\kappa_n$ with the principal curvatures.

The condition for principal directions is (A.7.0), namely

$$\mathbf{n}.d\mathbf{n} \times d\mathbf{r} = 0.$$

Let

$$\alpha \equiv \frac{FM - GL}{H^2}, \quad \beta \equiv \frac{FL - EM}{H^2}, \quad \gamma \equiv \frac{FM - EN}{H^2}, \quad \delta \equiv \frac{FN - GM}{H^2},$$

then

$$d\mathbf{n} = (\alpha\mathbf{r}_u + \beta\mathbf{r}_v)\,du + (\delta\mathbf{r}_u + \gamma\mathbf{r}_v)\,dv,$$

and

$$\mathbf{dr} = \mathbf{r}_u \mathrm{d}u + \mathbf{r}_v \mathrm{d}v.$$

Now $\mathbf{r}_u . \mathbf{r}_v = H\mathbf{n}$ from (1.1.5).
Hence

$$\mathbf{dn} . \mathbf{dr} = \alpha H\mathbf{n} . \mathrm{d}u\, \mathrm{d}v - \beta H\mathbf{n}\, \mathrm{d}u^2 + \delta H\mathbf{n}\, \mathrm{d}v^2 - \gamma H\mathbf{n} . \mathrm{d}u . \mathrm{d}v,$$

$$\mathbf{n} . \mathbf{dn} . \mathbf{dr} = H[-\beta\, \mathrm{d}u^2 + (\alpha - \gamma)\, \mathrm{d}u\, \mathrm{d}v - \delta\, \mathrm{d}v^2] = 0.$$

Hence

$$(FL - EM)\left(\frac{\mathrm{d}u}{\mathrm{d}v}\right)^2 - (\cancel{FM} - GL - \cancel{FM} + EN)\left(\frac{\mathrm{d}u}{\mathrm{d}v}\right) + (FN - GM) = 0,$$

which is equation (A.8.1). Thus the principal directions are extremal.

The following properties of normal curvature $\kappa_n$ have now been established. Through every point, P, of a smooth surface the curvature in general varies with the direction through the point. There are two orthogonal directions on which the $\kappa_n$ is extremal, and these are what have been named the principal directions. Their defining property is that adjacent normals at a pair of neighbouring points on a principal direction are co-planar.

For all other directions in general the adjacent normals are skew lines. As a result the surface along these directions has *twist* as well as (normal) curvature. Only along the two orthogonal principal directions is the twist zero.

*Exercise:* Show that the extreme values of $\kappa_n$ are given by

$$(\kappa_n)_e = \frac{M + N\lambda}{F + G\lambda} = \frac{L + M\lambda}{E + F\lambda},$$

where $\lambda$ is a root of (A.8.1).

Thus far the principal directions have not been thought of as being the coordinate directions. Indeed they will only be the coordinate directions if specially chosen. First, however, if the coordinate directions are principal then $F \equiv 0$, since the directions must be orthogonal. In the next section it will be shown that the further requirement that $M \equiv 0$ should also be true.

## A.9 Curvature and twist along the coordinate lines

From the available equations, but especially the Weingarten equations (A.5.1), all the required information relating to curvature and twist can be derived.

Consider a typical point P and the u coordinate direction passing through this point. Along this direction u is changing but v is not. This is the defining property. Then the curvature, $\kappa_{uu}$, of the u direction is defined as the negative arc rate of change of **n** in the u direction. This definition can be shown to be equivalent to that proposed earlier (A.1.0, A.1.2, A.4.3).

Expressed in symbols this definition becomes

$$\kappa_{uu} = -\frac{\mathbf{r}_u}{\sqrt{E}} . \mathbf{n}_s = -\frac{\mathbf{r}_u . \mathbf{n}_u}{E},$$

$$= -\left(\frac{(FM-GL)}{H^2}\right) . \frac{E}{E} - \left(\frac{(FL-EM)}{H^2}\right) . \frac{F}{E},$$

$$= \frac{L}{E}. \qquad \text{(A.9.0)}$$

If instead of the u direction, the v direction is considered then the curvature, $\kappa_{vv}$, of the v line will be found to be

$$\kappa_{vv} = \frac{N}{G}. \qquad \text{(A.9.1)}$$

*Worked example:* Consider the sphere referred to a cartesian coordinate system (Fig. A.7) and evaluate the curvatures $\kappa_{xx}$, $\kappa_{yy}$.

Now the equation of the sphere referred to these coordinates is

$$x^2 + y^2 + (z-R)^2 = R^2,$$

or

$$z = R - \sqrt{R^2 - (x^2 + y^2)}.$$

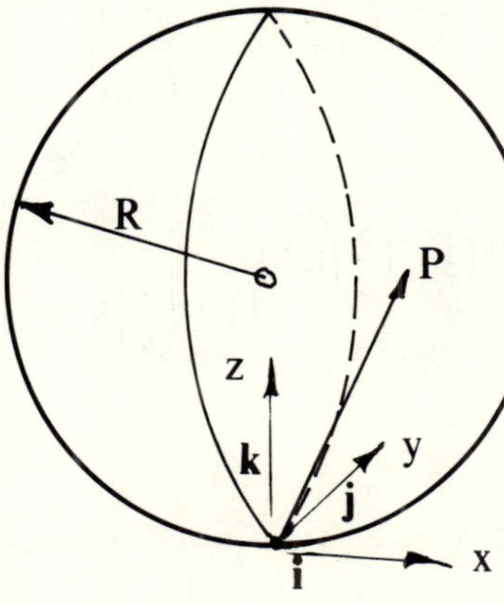


**Figure A.7** Geometry on the sphere.

Hence at a general point P,

$$\mathbf{r} = x\mathbf{i}+y\mathbf{j}+z\mathbf{k} = x\mathbf{i}+y\mathbf{j}+(R-\alpha)\mathbf{k},$$

where

$$\alpha \equiv \sqrt{R^2-(x^2+y^2)} = f(x,y).$$

Then

$$\mathbf{r}_x = \mathbf{i}+\frac{x}{\alpha}.\mathbf{k}, \quad \mathbf{r}_y = \mathbf{j}+\frac{y}{\alpha}.\mathbf{k}.$$

Hence

$$E \equiv \mathbf{r}_x.\mathbf{r}_x = 1+\left(\frac{x}{\alpha}\right)^2, \quad F \equiv \mathbf{r}_x.\mathbf{r}_y = \frac{xy}{\alpha^2}, \quad G \equiv \mathbf{r}_y\mathbf{r}_y = 1+\left(\frac{y}{\alpha}\right)^2.$$

Thus

$$\mathbf{n} = \frac{\mathbf{r}_x \times \mathbf{r}_y}{H} = \frac{-(x\mathbf{i}+y\mathbf{j})+\alpha\mathbf{k}}{R},$$

where

$$H^2 = 1+\frac{x^2+y^2}{\alpha^2}\,(=EG-F^2) = \left(\frac{R}{\alpha}\right)^2.$$

Then the curvatures $\kappa_{xx}$, $\kappa_{yy}$ can be evaluated as

$$L = -\mathbf{n}_x.\mathbf{r}_x = \frac{\mathbf{i}+\left(\frac{x}{\alpha}\right)\mathbf{k}}{R}.\left(\mathbf{i}+\left(\frac{x}{\alpha}\right)\mathbf{k}\right) = \frac{1}{R}\left(1+\left(\frac{x}{\alpha}\right)^2\right)$$

and

$$\kappa_{xx} = L/E = \frac{1}{R}\frac{\left(1+\left(\frac{x}{\alpha}\right)^2\right)}{\left(1+\left(\frac{x}{\alpha}\right)^2\right)} = \kappa(=\text{const.}).$$

In a similar manner

$$N = -\mathbf{n}_y.\mathbf{r}_y = \frac{\left(\mathbf{j}+\left(\frac{y}{\alpha}\right)\mathbf{k}\right)}{R}.\left(\mathbf{j}+\left(\frac{y}{\alpha}\right)\mathbf{k}\right) = \kappa\left(1+\left(\frac{y}{\alpha}\right)^2\right)$$

and

$$\kappa_{yy} = N/G = \frac{\kappa\left(1+\left(\frac{y}{\alpha}\right)^2\right)}{\left(1+\left(\frac{y}{\alpha}\right)^2\right)} = \kappa(=\text{const.}).$$

Hence the two curvatures are equal, and each is equal to the (constant) curvature of the sphere—as might have been anticipated.

In addition to the curvatures $\kappa_{uu}$, $\kappa_{vv}$, the directions u, v in general have associated twists which are also needed for a complete discussion of curvature at a typical point P. The twist at P in the u direction is defined as the negative arc rate of change of **n** *normal* to u. By use of (A.5.1) the appropriate component for twist about u can be seen to be

$$\kappa_{uv} = -\left(\frac{FL-EM}{H^2}\right)\sqrt{G}\,.\frac{H}{\sqrt{EG}}.\frac{1}{\sqrt{E}},$$

where now the second term in the $\mathbf{n}_u$ expression has been multiplied by $\sqrt{G}$, the magnitude of $\mathbf{r}_v$, then divided by $\sqrt{E}$ to give the arc rate of change and multiplied by $\sin\theta = H/\sqrt{EG}$ to give the relevant component normal to $\mathbf{r}_u$. After simplification there is obtained

$$\kappa_{uv} = \left(\frac{EM-FL}{EH}\right). \qquad (A.9.2)$$

By a similar argument, the twist along v ($\kappa_{vu}$) is given by

$$\kappa_{vu} = \left(\frac{GM-FN}{GH}\right). \qquad (A.9.3)$$

These twist components can alternatively be shown to represent the arc rates at which n *rotates* about each of u and v, for changes in u and v respectively.

Note that if the coordinate lines (u, v) are orthogonal then F = 0 and

$$\kappa_{uv} = \kappa_{vu} = M/\sqrt{EG}. \qquad (A.9.4)$$

Generally, however, for *non-orthogonal* u, v lines, the two twists are not equal. Exceptionally, $\kappa_{uv} = \kappa_{vu}$ also, should $L = N = 0$. Such surfaces and choice of u, v do occur but are not common.

*Exercise:* The earlier exercise in this section can now be further examined, for twists. Using the results found there,

$$M \equiv -\mathbf{n}_x.\mathbf{r}_y = +\kappa\left(\mathbf{i}+\left(\frac{x}{\alpha}\right)\mathbf{k}\right).\left(\mathbf{j}+\left(\frac{y}{\alpha}\right)\mathbf{k}\right)$$

$$= +\kappa\frac{xy}{\alpha^2}.$$

Hence

$$\kappa_{uv} = \left(\frac{EM-FL}{EH}\right) = \frac{\left[1+\left(\frac{x}{\alpha}\right)^2\right]\times\kappa\frac{xy}{\alpha^2}-\frac{xy}{\alpha^2}\kappa\left(1+\left(\frac{x}{\alpha}\right)^2\right)}{\left[1+\left(\frac{x}{\alpha}\right)^2\right].\kappa/\alpha}$$

$$= 0 = \kappa_{vu}.$$

Hence there are no twists associated with these directions x, y when projected onto the sphere. This result too perhaps could have been anticipated, because of the special nature of the sphere, where every direction is a principal direction.

The defining property for principal directions has been expressed in terms of the normals of adjacent points along the particular direction being *co-planar*. This is equivalent to the requirement that the *twists* should be zero along principal directions.

Now, as noted in A.8, the principal directions are certainly orthogonal and hence $F \equiv 0$. If in addition the associated twists along these directions should be zero, then from (A.9.4), it is seen that this requirement is met if $M \equiv 0$.

Hence the necessary (and sufficient) condition for the *coordinate directions* to be *principal directions* is that

$$\boxed{F \equiv M \equiv 0} \qquad \text{(A.9.5)}$$

*Exercise:* The case of the sphere treated earlier in this section produced $\kappa_{uv} = 0$ for all directions. This is a special case, where *all* directions are principal, even if the two coordinates chosen are not orthogonal.

## A.10 The curvature matrix

In the previous section the curvatures and twists along the u, v coordinate directions were evaluated. In use the four quantities are closely associated; indeed, it is most useful to think of the quantities $\kappa_{uu}$, $\kappa_{vv}$, $\kappa_{uv}$ and $\kappa_{vu}$ as component parts of a single quantity, the curvature matrix $\boldsymbol{\kappa}$, which is defined as

$$\boldsymbol{\kappa} \equiv \begin{bmatrix} \kappa_{uu} & \kappa_{uv} \\ \kappa_{vu} & \kappa_{vv} \end{bmatrix}, \qquad \text{(A.10.0)}$$

$$= \begin{bmatrix} \frac{L}{E} & \left(\frac{EM-FL}{EH}\right) \\ \left(\frac{GM-FN}{GH}\right) & \frac{N}{G} \end{bmatrix}.$$

When the coordinates are general (and $F \neq 0$) this is a non-symmetric matrix. If the coordinates are orthogonal, then $F = 0$ and the matrix becomes symmetric with an off-diagonal term of $M/\sqrt{EG}$. In other words, the curvature matrix for orthogonal coordinates becomes

$$\boldsymbol{\kappa} = \begin{bmatrix} \dfrac{L}{E} & \dfrac{M}{\sqrt{EG}} \\ \dfrac{M}{\sqrt{EG}} & \dfrac{N}{G} \end{bmatrix}. \qquad \text{(A.10.1)}$$

The elements in this matrix are the curvatures and twist along two orthogonal directions, u, v, the coordinate directions. If $M \neq 0$ then these directions, though orthogonal, are not principal. Since the coordinate directions u, v are chosen at the outset, usually for convenience, and the principal curvature directions on the deformed plate surface will depend upon how the plate is loaded, it is seldom the case that the u, v directions are principal curvature directions.

Consider an orthogonal choice of coordinates. The question then arises, how can the principal curvature directions through a typical point P be identified, once E, G; L, M, N are known? Now $F = 0$, since the coordinates (u, v) have been chosen to be orthogonal. From (A. 8.1), when $F = 0$, the directions for the principal curvatures are given by the roots of the quadratic

$$EM\,.\,\lambda^2 + (EN - GL)\lambda - GM = 0. \qquad \text{(A.10.2)}$$

Hence these directions can be formed, and so too can the principal curvatures from (A.7.3) as the roots of

$$EG\kappa_p^2 - \kappa_p(EN + GL) + T^2 = 0. \qquad \text{(A.10.3)}$$

If $F \neq 0$, then the general equations (A.7.3) and (A.8.1) must be used.

*Exercise:* For the H. P., $z = Kxy$, show that when (x, y) are the chosen coordinates then the curvature matrix has the form

$$\boldsymbol{\kappa} = \begin{bmatrix} 0 & \dfrac{K}{H^2} \\ \dfrac{K}{H^2} & 0 \end{bmatrix},$$

i.e. the curvatures are zero, and the twist, in either sense, xy or yx, has the value $K/H^2$. Thus, although the coordinate lines are non-orthogonal, the curvature matrix is symmetric, but here because of the special circumstance that $L = N = 0$.

## A.11 The curvature circle

The information summarized by the curvature matrix (A.10.0) can for some purposes be more readily represented graphically on what will be called the *curvature circle.*

First recall that at every point, P, on every smooth surface there are just two directions, the principal directions, along which the twist is zero. These directions when extended to adjacent points form a possible system of orthogonal coordinate lines on the surface, called the *lines of curvature* on the surface. Referred to these lines as coordinates, for example lines of latitude and longitude on a perfectly spherical globe, then the curvature matrix becomes a diagonal matrix, since the twist is zero. As noted in A.9, for this choice F = M = 0.

Generally, however, for coordinates which are not lines of curvature, then the curvature matrix will be as shown in (A.10.0). The matrix will be *symmetric* if the coordinates chosen are *orthogonal,* though not necessarily principal.

The curvature circle is drawn on axes of curvature and twist. The circle is centred on the curvature axis (Fig. A.8).

Consider first the case when the coordinate lines (u, v) are orthogonal, though not necessarily lines of curvature. If the coefficients of the fundamental forms are E, G; L, M, N, then

$$\kappa_{uu} = \frac{L}{E}, \quad \kappa_{uv} = \frac{M}{\sqrt{EG}} = \kappa_{vu}, \quad \kappa_{vv} = \frac{N}{G}. \tag{A.11.0}$$

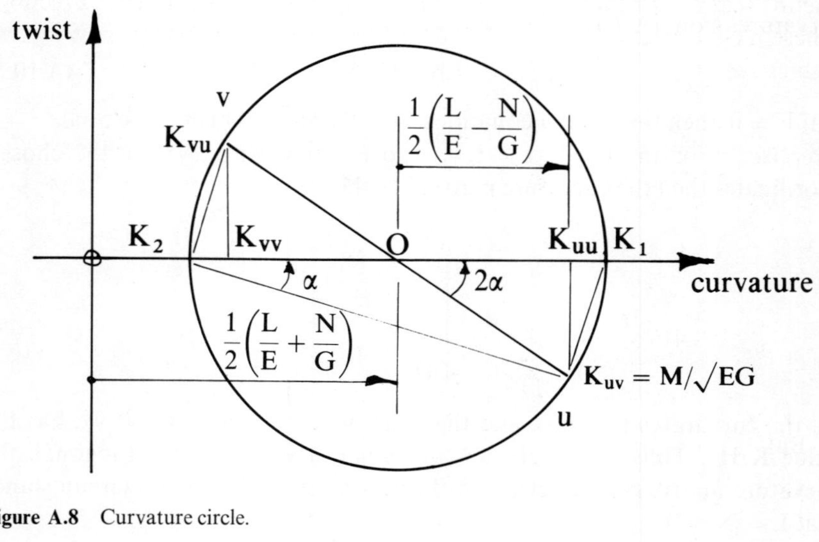


**Figure A.8** Curvature circle.

Now from (A.10.3) the *principal curvatures* are given by the roots of the quadratic

$$EG\kappa_p^2 - \kappa_p(EN+GL)+(LN-M^2) = 0.$$

This can be rewirtten as

$$\kappa_p^2 - \kappa_p\left(\frac{L}{E}+\frac{N}{G}\right)+\left(\frac{L}{E}.\frac{N}{G}-\left(\frac{M}{\sqrt{EG}}\right)^2\right) = 0.$$

Hence, if the principal values are $\kappa_1, \kappa_2$ then

$$\kappa_1+\kappa_2 = \frac{L}{E}+\frac{N}{G}, \tag{A.11.1}$$

and

$$\kappa_1 . \kappa_2 = \frac{L}{E}.\frac{N}{G} - \left(\frac{M}{\sqrt{EG}}\right)^2.$$

Let

$$4R^2 \equiv \left(\frac{L}{E}-\frac{N}{G}\right)^2 + 4\left(\frac{M}{\sqrt{EG}}\right)^2.$$

Then

$$\kappa_{1,2} = \frac{1}{2}\left(\frac{L}{E}+\frac{N}{G}\right) \pm R. \tag{A.11.2}$$

The relationships (A.11.1) and (A.11.2) are the basis for the circle construction (Fig. A.8). In addition to the principal values it is important to find the inclination of the principal directions to the (u, v) coordinate directions. These inclinations are obtained from (A.10.2) where $\lambda = du/dv$, and the two values are

$$\lambda_{1,2} = \frac{(GL-EN)\pm\sqrt{(GL-EN)^2-4.(EM)(-GM)}}{2EM},$$

$$= \frac{\frac{1}{2}\left(\frac{L}{E}-\frac{N}{G}\right)\pm\sqrt{\frac{1}{4}\left(\frac{L}{E}-\frac{N}{G}\right)^2+\left(\frac{M}{\sqrt{EG}}\right)^2}}{M/G},$$

$$= \frac{\frac{1}{2}\left(\frac{L}{E}-\frac{N}{G}\right)\pm R}{M/G}.$$

But

$$\tan\alpha = \sqrt{\frac{G}{E}}.\frac{dv}{du} = \sqrt{\frac{G}{E}}.\frac{1}{\lambda},$$

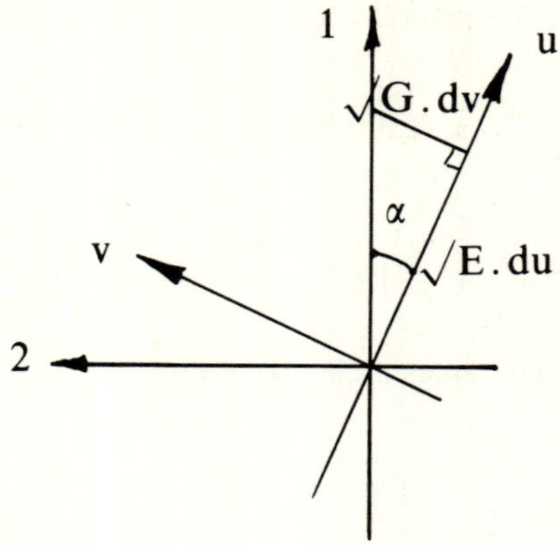


**Figure A.9** Principal directions.

from Fig. A.9

$$= \frac{\dfrac{M}{\sqrt{EG}}}{\dfrac{1}{2}\left(\dfrac{L}{E} - \dfrac{N}{G}\right) \pm R}.$$

The plus sign in this expression is associated with the angle shown in Fig. A.8. The minus sign is associated with an angle $[-(\pi/2-\alpha)]$. On the curvature circle it will be noted that with respect to the circle centre, O, an angle of $2\alpha$ is turned through clockwise to arrive at the first principal direction, 1. Alternatively, if an angle $2[(\pi/2)-\alpha]$ from the u direction is turned counter-clockwise then the second principal direction, 2, is obtained.

Thus *directions* in the physical tangent plane of the surface, such as u, v; 1, 2 are represented by *points* on the curvature circle. Further, with centre O, in the circle, angles *twice* the *physical* angles between particular *directions* must be turned to obtain the corresponding circle *point*. There is a direct analogy between the curvature circle and the Mohr circle of stress or strain.

In the general case of non-orthogonal coordinates, then the curvature circle is drawn in a slightly different manner. The formulae are now more complicated because there is no symmetry of the curvature matrix, and $F \neq 0$. The circle appears as in Fig. A.10.

The following relations are implied by the circle geometry.

$$\beta+\theta = \frac{\pi}{2}+\alpha, \quad \text{hence} \quad \cot\alpha = -\tan(\beta+\theta). \tag{A.11.3}$$

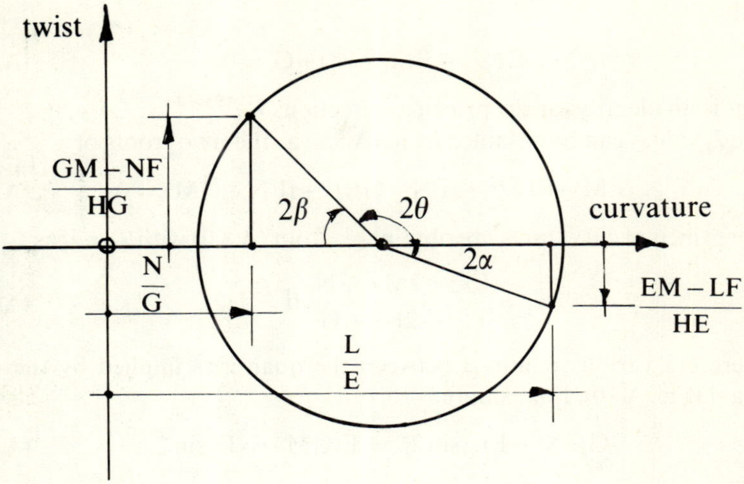


**Figure A.10** Curvature circle—non-orthogonal coordinates.

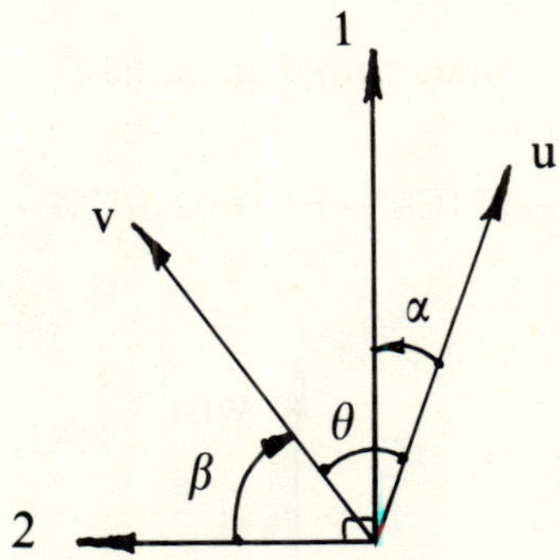


**Figure A.11** Principal directions—non-orthogonal coordinates.

If the principal directions are given by the two values of

$$\lambda_i = \frac{d\alpha}{d\beta}\ (i = 1, 2), \quad \text{then}$$

$$\cot\alpha = \left(\frac{E\lambda_1 + F}{H}\right) \quad \text{and} \quad \cot\beta = -\left(\frac{G + F\lambda_2}{H\lambda_2}\right). \tag{A.11.4}$$

Now

$$\operatorname{Tan}\theta = \frac{H}{F} \quad \text{and hence} \quad \operatorname{Tan}(\theta + \beta) = \frac{H}{E\lambda_2 + F}. \tag{A.11.5}$$

Finally,

$$E\lambda_1\lambda_2 + F(\lambda_1+\lambda_2) + G = 0. \tag{A.11.6}$$

This is an identity for the principal directions $\lambda_1$, $\lambda_2$.

The $\lambda_i$ values can be obtained from (A.8.1) as the (two) roots of

$$(EM-FL)\lambda^2 + (EN-GL)\lambda + (FN-GM) = 0. \tag{A.11.7}$$

The principal curvatures are obtainable from (A.8.0) with $\lambda = \lambda_i$ as

$$\kappa_i = \frac{L\lambda_i^2 + 2M\lambda_i + N}{E\lambda_i^2 + 2F\lambda_i + G}, (i = 1, 2). \tag{A.11.8}$$

There are various relations between the quantities implied by the circle diagram (Fig. A.10). For example

$$G(EM-FL)\sin 2\beta = E(GM-NF)\sin 2\alpha. \tag{A.11.9}$$

Also

$$\operatorname{Sin} 2\theta = \frac{2HF}{EG}, \quad \operatorname{Cos} 2\theta = \frac{2F^2-EG}{EG}, \tag{A.11.10}$$

from the definitions of the fundamental quantities.

Another useful relation is

$$E(LG-EN) + 2F(EM-FL) = 2H(EM-FL)\operatorname{Cot} 2\alpha. \tag{A.11.11}$$

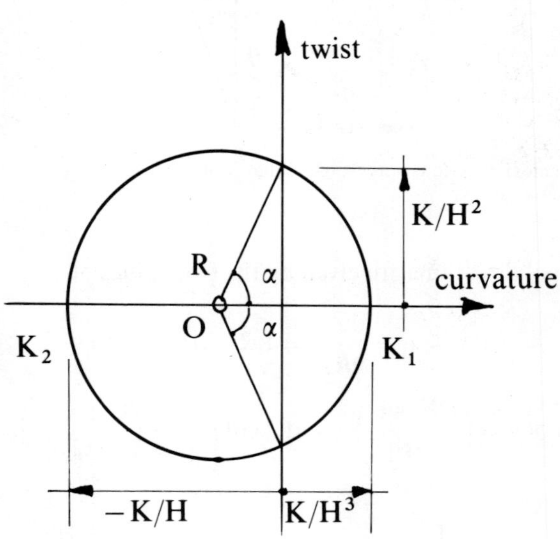


**Figure A.12** Curvature circle—hyperbolic paraboloid.

The basic results for the coordinate directions u, v are of course the relations

$$\kappa_{uu} = \frac{L}{E}, \quad \kappa_{uv} = \frac{(EM-FL)}{EH}, \quad \kappa_{vu} = \frac{(GM-FN)}{GH}, \quad \kappa_{vv} = \frac{N}{G}. \qquad (A.11.12)$$

These are the component values occurring in the curvature matrix.
*Exercise:* For the H. P. $z = Kxy$, with $(x, y)$ chosen as coordinates, show that the curvature circle is given by Fig. A.12.

## A.12 Continuity requirements

The core of the study of surface geometry as presented so far depends upon the calculation of six quantities in general, namely E, F, G; L, M, N, the coefficients of the two fundamental forms.

It might be thought that these six quantities are independent; however, they are not. There exist *three* independent relations between the *six* quantities. These relations (the *continuity* requirements) are differential in form and are the requirements that the six quantities should describe a *smooth* surface.

Collectively the continuity requirements are known as the *Codazzi–Gauss equations*; there are two Codazzi equations and one Gauss equation, three equations in all. Should there be some special features about the particular surface and choice of coordinates, then some of the fundamental quantities may be zero and some of the continuity conditions may be identities.

Initially, suppose that the coordinates u, v are *line of curvature* coordinates. Then $F = M \equiv 0$. For this choice of coordinates, it is useful to define $A^2 \equiv E$, $B^2 \equiv G$. The A and B are known as the *Lamé parameters*. There are now just four fundamental quantities A, B; L, N.

The three vectors $\mathbf{r}_u$, $\mathbf{r}_v$, $\mathbf{n}$ in this case form a mutually perpendicular system, with $\mathbf{n}$ a unit vector, the normal vector. For definiteness we shall assume $\mathbf{r}_n$, $\mathbf{r}_v$, $\mathbf{n}$ to be a right-handed system. Now evaluate the second derivatives of $\mathbf{r}$ in terms of these three (base) vectors.

From the definitions of L and N, and remembering that M is zero for line of curvature coordinates, then

$$\begin{aligned} \mathbf{r}_{uu} &= L\mathbf{n} + \alpha . \mathbf{r}_u + a . \mathbf{r}_v \\ \mathbf{r}_{uv} &= 0 . \mathbf{n} + \beta . \mathbf{r}_u + b . \mathbf{r}_v \\ \mathbf{r}_{vv} &= N\mathbf{n} + \gamma \mathbf{r}_u + c . \mathbf{r}_v \end{aligned} \qquad (A.12.0)$$

where $\alpha, \beta, \gamma$; a, b, c are coefficients, to be evaluated.

Now

$$\mathbf{r}_u \cdot \mathbf{r}_{uu} = \tfrac{1}{2}(\mathbf{r}_u^2)_u = \tfrac{1}{2}E_u.$$

Also

$$\mathbf{r}_v \cdot \mathbf{r}_{uu} = (\mathbf{r}_u \cdot \mathbf{r}_v)_u - \tfrac{1}{2}(\mathbf{r}_u^2)_v = O - \tfrac{1}{2}E_v,$$

together with the similar expressions, $\mathbf{r}_u \cdot \mathbf{r}_{uv} = \frac{1}{2}E_v$,

$$\mathbf{r}_v \cdot \mathbf{r}_{uv} = \tfrac{1}{2}G_u; \quad \mathbf{r}_u \cdot \mathbf{r}_{vv} = -\tfrac{1}{2}G_u \quad \text{and} \quad \mathbf{r}_v \cdot \mathbf{r}_{vv} = \tfrac{1}{2}G_v.$$

Then by taking $\mathbf{r}_u$., $\mathbf{r}_v$. in turn of the expressions in (A.12.0) the unknown coefficients $\alpha, \beta, \gamma$; a, b, c can be found.

Thus

$$\mathbf{r}_u \cdot \mathbf{r}_{uu} = \frac{1}{2}E_u = \alpha \cdot E, \quad \text{hence} \quad \alpha = \frac{1}{2E} \cdot E_u,$$

$$\mathbf{r}_v \cdot \mathbf{r}_{uu} = -\frac{1}{2}E_v = aG, \quad \text{and} \quad a = -\frac{1}{2G}E_v. \qquad \text{(A.12.1)}$$

Similarly,

$$\beta = \frac{1}{2E}E_v, \ b = \frac{1}{2G}G_u; \quad \gamma = \frac{1}{2E}G_u, \ c = \frac{1}{2G}G_v.$$

Consider now the identity

$$(\mathbf{r}_{uu})_v = (\mathbf{r}_{uv})_u.$$

From the knowledge of $\mathbf{r}_{uu}$, $\mathbf{r}_{uv}$ in terms of $\mathbf{r}_u$, $\mathbf{r}_v$, $\mathbf{n}$, this identity can be written in expanded form as

$$L_v\mathbf{n} + L\mathbf{n}_v + \alpha_v\mathbf{r}_u + \alpha\mathbf{r}_{uv} + a_v\mathbf{r}_v + a\mathbf{r}_{uv} = \beta_u\mathbf{r}_u + \beta\mathbf{r}_{uu} + b_u\mathbf{r}_v + b\mathbf{r}_{uv}. \qquad \text{(A.12.2)}$$

If the expressions for the second derivatives of $\mathbf{r}$ from (A.12.0) are substituted, and the coefficient of $\mathbf{n}$ in the resulting expression with a zero right-hand side computed, then this must be zero. So too must the coefficients of $\mathbf{r}_u$ and $\mathbf{r}_v$ be separately zero.

From (A.12.2)

$$(L_v - L\beta + aN)\mathbf{n} + (\ )\mathbf{r}_u + (\ )\mathbf{r}_v = 0.$$

Hence

$$L_v - \beta . L + aN = 0.$$

But

$$L = E\kappa_1, \quad N = G\kappa_2, \quad E = A^2, \quad G = B^2,$$

whence

$$(A^2\kappa_1)_v - \frac{A^2\kappa_1}{2A^2}(A^2)_v = \frac{1}{2B^2}(A^2)_v B^2\kappa_2.$$

Simplifying,

$$\boxed{(A\kappa_1)_v = \kappa_2(A)_v} . \tag{A.12.3a}$$

In a similar manner, by equating the **n** component of $(\mathbf{r}_{vv})_u = (\mathbf{r}_{uv})_v$ to zero there is obtained the equation,

$$\boxed{(B\kappa_2)_u = \kappa_1(B)_u} . \tag{A.12.3b}$$

These two continuity requirements are known as the Codazzi equations. The relations which follow from equating to zero any of the other four components of the vector identities can be shown to give rise to just a single further equation of continuity, the Gauss equation. (The $\mathbf{r}_u$ component of (A.12.2) set to zero leads to an identity, but the $\mathbf{r}_v$ component equated to zero leads to the Gauss equation).

Proceed then to equate to zero the $\mathbf{r}_v$ component of (A.12.2), when there is obtained

$$-\kappa_2 L + \left(\frac{1}{2E}E_u\right)\cdot\left(\frac{1}{2G}G_u\right) + \left(-\frac{1}{2G}E_v\right)_v + \left(-\frac{1}{2G}E_v\right)\cdot\left(\frac{1}{2G}G_v\right)$$
$$= \left(\frac{1}{2E}E_v\right)\cdot\left(-\frac{1}{2G}E_v\right) + \left(\frac{1}{2G}G_u\right)_u + \left(\frac{1}{2G}G_u\right)^2.$$

This equation can be simplified. Now $L = E\kappa_1$, $E = A^2$, $G = B^2$, then there results

$$\boxed{\left(\frac{1}{A}B_u\right)_u + \left(\frac{1}{B}A_v\right)_v = -AB\kappa_1\kappa_2.} \tag{A.12.4}$$

This equation of continuity is known as the Gauss equation.

The importance of the Gauss (continuity) equation is that it shows the Gaussian curvature $\kappa_g(\equiv\kappa_1\kappa_2)$ to be a function of A and B only. Note that L, N do not appear in (A.12.4). This is a somewhat surprising, but very important result. Thus it shows that any surface which is *deformed* in such a way that the surface is not stretched, sheared or compressed undergoes *no change* of $\kappa_g$ at any point, since under these conditions of *inextensional bending* of the surface, neither A nor B for the surface change at any point, and hence the expression for $\kappa_g$ implied by (A.12.4) remains unchanged.

In addition to the Codazzi–Gauss continuity equations, in some chapters other continuity considerations will arise from the need to ensure slope continuity between adjacent portions of surface. Particularly in the study of optimized plates, there is need to discuss (deflection) surfaces which are made up from portions of different types of surface. A requirement will be that such surfaces should be fully continuous within each portion, that is show *gradual* variation of all quantities such as coordinates, slope and curvature. A composite surface made up from a number of such portions must in addition be coordinate and *slope* continuous across and along the common junction. This requires that pairs of coincident points on the junction, regarded as belonging each to an adjacent portion, should possess common coordinates and common tangent plane. When dealing with plates, which when deflected by load are *shallow* surfaces (see p. 177), such conditions will usually be expressed in terms of displacements and surface slopes.

The details of such problems are dealt with in chapter 5.

## A.13 Special surfaces

Thus far the discussion has considered general surfaces. It is appropriate now to turn to various special classes of surface.

Consider first the class of *surfaces of revolution*. Such surfaces are traced out when a plane curve, coplanar and fixed with respect to an axis, is rotated about this axis. The axis will be referred to as the axis of revolution. If this is taken as the axis for a coordinate, z and d denotes the perpendicular distance to the curve, then the coordinates of a point can be written as

$$x = d(z)\cos\phi, \quad y = d(z)\sin\phi, \quad z. \tag{A.13.0}$$

Here x, y are axes fixed in space with an origin on the axis of revolution (z), and $\phi$ is a longitude angle seen true when the surface is viewed along the z axis. The surface coordinates, u, v will be chosen so that $u(=z)$ varies in the axial direction, and $v(=\phi)$ varies in the circumferential direction.

Then the position vector of a typical point is given by

$$\mathbf{r} = d\cos\phi\,.\,\mathbf{i} + d\sin\phi\,.\,\mathbf{j} + z\mathbf{k} \tag{A.13.1}$$

where **i**, **j**, **k** are fixed unit vectors along x, y, z respectively. The surface tangent vectors $\mathbf{r}_z, \mathbf{r}_\phi$ can then be found as

$$\begin{aligned}\mathbf{r}_z &= dz\cos\phi\,.\,\mathbf{i} + dz\sin\phi\,.\,\mathbf{j} + \mathbf{k}\\ \mathbf{r}_\phi &= -d\sin\phi\,.\,\mathbf{i} + d\cos\phi\,.\,\mathbf{j} + 0\end{aligned} \tag{A.13.2}$$

Here

$$(\ )z = \frac{\partial(\ )}{\partial z}, \quad (\ )_\phi = \frac{\partial(\ )}{\partial\phi}.$$

The coefficients of the first fundamental form are

$$\begin{aligned}
E &= \mathbf{r}_z \cdot \mathbf{r}_z = dz^2 \cos^2 \phi + dz^2 \sin^2 \phi + 1,\\
&= 1 + dz^2,\\
F &= \mathbf{r}_z \cdot \mathbf{r}_\phi = -dz \cos \phi \,.\, \sin \phi + dz \sin \phi \,.\, \cos \phi = 0,\\
G &= \mathbf{r}_\phi \cdot \mathbf{r}_\phi = d^2.
\end{aligned} \tag{A.13.3}$$

From these results it is seen that the coordinate lines are orthogonal, since $F \equiv 0$. Hence $\mathbf{r}_z$ and $\mathbf{r}_\phi$ are orthogonal.

Next the unit normal can be evaluated as

$$\mathbf{n} = \frac{\mathbf{r}_z \times \mathbf{r}_\phi}{(\mathbf{r}_z \times \mathbf{r}_\phi)} = \frac{-d \cos \phi \,.\, \mathbf{i} - d \sin \phi \,.\, \mathbf{j} + d \,.\, d_z \mathbf{k}}{d\sqrt{(1+dz^2)}}.$$

The second derivatives of $\mathbf{r}$ are

$$\begin{aligned}
\mathbf{r}_{zz} &= d_{zz} \cos \phi \,.\, \mathbf{i} + d_{zz} \sin \phi \,.\, \mathbf{j} + 0,\\
\mathbf{r}_{z\phi} &= dz \sin \phi \,.\, \mathbf{i} + dz \cos \phi \,.\, \mathbf{j} + 0,\\
\mathbf{r}_{\phi\phi} &= -d \cos \phi \,.\, \mathbf{i} - d \,.\, \sin \phi \,.\,.\, \mathbf{j} + 0.
\end{aligned}$$

Here the coefficients of the second fundamental form are

$$\begin{aligned}
L &\equiv \mathbf{n} \,.\, \mathbf{r}_{zz} = \frac{-d_{zz}}{\sqrt{(1+dz^2)}},\\
M &\equiv \mathbf{n} \,.\, \mathbf{r}_{z\phi} = 0,\\
N &\equiv \mathbf{n} \,.\, \mathbf{r}_{\phi\phi} = \frac{d}{\sqrt{(1+dz^2)}}.
\end{aligned} \tag{A.13.4}$$

Since $M \equiv 0$, and $F = 0$, it follows that the coordinate curves on the surface associated with z, $\phi$ are lines of curvature.

Hence the principal curvatures are given by

$$\kappa_1 = L/E = -\frac{d_{zz}}{(1+dz^2)^{3/2}} \tag{A.13.5}$$

and

$$\kappa_2 = N/G = \frac{1}{d(1+dz^2)^{1/2}}.$$

These expressions can be interpreted as follows.

From Fig. (A.13) it can be seen that $\kappa_1 = 1/O_1P$, where $O_1$ is the centre of curvature for the original plane curve, at the point P. The minus sign relates to $\mathbf{n}$ being directed *away* from $O_1$. The point $O_2$ where the normal intersects the axis of revolution is the centre of curvature associated with the $\phi$ coordinate direction since from the geometry of the figure, it can be seen that

$$\kappa_2 = \frac{1}{O_2P}.$$

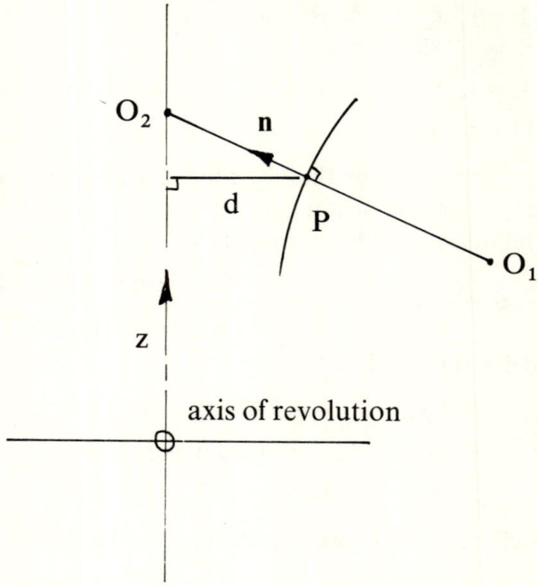


**Figure A.13** Surface of revolution—section.

Here there is a *positive* sign since **n** and $O_2$ are on the *same* side of the surface.

As a point of interest, for Codazzi–Gauss continuity conditions it can be noted that

$$A = \sqrt{1+d_z^2}, \quad B = d; \quad u = z, \quad v = \phi, \quad d = d(z).$$

Hence the Gauss equation becomes

$$\left(\frac{1}{A}B_z\right)_z + \left(\frac{1}{B}A_\phi\right)_\phi = -AB\kappa_1\kappa_2.$$

Now

$$\text{L.H.S.} = \left(\frac{1}{\sqrt{1+d_z^2}}\,.\,dz\right)_z + 0$$

$$= \frac{\sqrt{1+dz^2}\,.\,d_{zz} - dz\,\dfrac{1\,.\,2dz\,.\,d_{zz}}{2\sqrt{1+d_z^2}}}{1+dz^2},$$

$$= \frac{d_{zz}[1+dz^2-dz^2]}{(1+dz^2)^{3/2}} = \frac{d_{zz}}{(1+d_z^2)^{3/2}}.$$

$$\text{R.H.S} = \frac{-\sqrt{1+dz^2}\,.\,d-d_{zz}}{(1+dz^2)^{3/2}}\cdot\frac{1}{d(1+dz^2)^{1/2}} = \text{L.H.S.}$$

One Codazzi equation only exists because of the symmetry and this becomes

$$(B\kappa_2)_u = \left(d\,.\,\frac{1}{d(1+dz^2)^{1/2}}\right)_z = -\frac{dz\,.\,d_{zz}}{(1+dz^2)^{3/2}} = \kappa_1(B)_u$$

an identity.

*Exercise:* The hyperboloid of revolution is defined by

$$\frac{x^2+y^2}{a^2}-\frac{z^2}{b^2}=1.$$

If

$$\Omega^2 \equiv b^2+z^2\left(1+\left(\frac{a}{b}\right)^2\right),$$

show that

$$\kappa_{11} = -b/\Omega a, \quad \kappa_{22} = ab/\Omega^3,$$

are the principal curvatures.

Another very important class of surfaces of interest is the class of *shallow surfaces.* In fact all the shapes developed by loaded plates, of the type concerned with in this book, are shallow surfaces. By this is meant that the surface *slopes* are everywhere *small.* As an example, this means that a shallow paraboloid of revolution cannot be distinguished from a shallow segmental cap cut from a sphere: in each case the shape locally is described by a quadratic variation.

Some surfaces can be thought of as traced out by the translation of one curve over another. Such surfaces are known collectively as *surfaces of translation.* If the two curves are straight lines, then an example of such a surface is the hyperbolic paraboloid, z = Kxy, which has been studied in the earlier exercises.

Certain surfaces have the property that by suitable *unfolding* they can be laid flat on the plane. These are the *developable surfaces.* Examples (amongst others) are cylinders and cones. But the sphere cannot be so unfolded, hence the need for some (agreed) projection (i.e. distortion) of the surface of the sphere earth in order to produce flat maps of a whole or part of the surface. The cylinders and cones when so unfolded are undergoing *inextensional bending*, the type of deformation which neither stretches, shears nor compresses any portion of the surface. As a result, the starting value of the Gaussian curvature, $\kappa_g$ (=product of the principal curvatures), is maintained. But since the ultimate unfolding on to the plane produces a surface (the plane) which clearly has zero $\kappa_g$, then so too must $\kappa_g$ be zero at all stages of the unfolding process.

Imagine a surface which spans on to a non-planar boundary. If we wish to know the shape of surface which spans to this boundary but has *minimum* surface area, then the search is for a *minimal* surface. Such a surface has the property that at every point the principal curvatures are equal and opposite.

Suppose there to be two circular wire hoops placed one above the other. The surface of revolution which has minimal area and spans to these two wire hoops is a catenoid—the surface obtained by revolving a catenary about the curve's own directrix. From the nature of surface tension, soap films are minimal surfaces when not subject to a pressure difference across the film. As an approximation, the hyperbolic paraboloid is a minimum surface, when the surface is *shallow*, namely when z is everywhere small compared with the plan dimensions of the surface.

There is one further class of surface of interest which has been discussed in some detail in chapter 5. This is the class of *constant curvature* surfaces. These surfaces are such that at all points on the surface there is *one principal curvature* of a *constant* value k, and the other principal curvature is numerically less than k.

## A.14 Summary—the geometrical quantities required for the construction of a plate theory

This chapter has been devoted to the study of surface geometry. It was pointed out at the beginning that surface geometry is important because all plates when loaded deform into the shape of some surface, albeit a *shallow* one. The properties of the resulting surface shape determine specifically all the mechanical quantities of interest in constructing a mechanics of plates.

The most important induced geometrical quantities in the deformed plate are the *curvatures* and *twists*. Hence the description and analysis of curvature and twist in this chapter is the central core of technique used in this book.

However other aspects of surface geometry are also important and have been used as and when required. These other aspects have come to the fore particularly when discussing, for example, constant curvature surfaces or finite elements. When particular examples expressed in terms of particular coordinate systems are discussed, the geometrical quantities such as curvatures and twists are computed using the methods set out in this chapter.

# Index